T0304745

Plant Growth Responses for Smart Agriculture

Plant Growth Responses for Smart Agriculture
Prospects and Applications

T. Girija
Professor and Head
Department of Plant Physiology
College of Horticulture
Kerala Agricultural University
Vellanikkara, KAU P.O.Thrissur, Kerala-680 656

Nandini K.
Department of Plant Physiology
College of Horticulture
Kerala Agricultural University
Vellanikkara, KAU P.O., Thrissur, Kerala-680 656

Parvathi M S
Department of Plant Physiology
College of Horticulture
Kerala Agricultural University
Vellanikkara, KAU P.O., Thrissur, Kerala-680 656

CRC Press is an imprint of the
Taylor & Francis Group, an **informa** business

NEW INDIA PUBLISHING AGENCY
New Delhi-110 034

First published 2022
by CRC Press
4 Park Square, Milton Park, Abingdon, Oxon, OX14 4RN

and by CRC Press
6000 Broken Sound Parkway NW, Suite 300, Boca Raton, FL 33487-2742

CRC Press is an imprint of Taylor & Francis Group, an Informa business

British Library Cataloguing-in-Publication Data
A catalogue record for this book is available from the British Library

Library of Congress Cataloging-in-Publication Data
A catalog record has been requested

ISBN: 978-1-032-15809-9 (hbk)
ISBN: 978-1-003-24574-2 (ebk)

DOI: 10.1201/9781003245742

Dedicated to
Our Parents and Teachers

Kerala Agricultural University

Main Campus, Kerala Agriculture University, P.O. Thrissur - 680 656, Kerala

Dr. R. Chandra Babu, Ph.D, Post Doc (USA)
Vice Chancellor

Foreword

Plant Physiology is a dynamic science which goes on adding knowledge to already characterized basic processes in plants. The past decade has witnessed an unprecedented progress in biological sciences with the advent of innovative technologies viz. recombinant DNA techniques, omics approaches and advanced phenotyping platforms. These tools have helped to redefine many of the already accepted facts of plant life. In this context, I am happy to present the publication **"Plant Growth Responses for Smart Agriculture: Prospects and Applications"** which gives an insight into the lesser known signals that can influence plant growth and development.

Knowledge of plant physiological processes provides the base for research in cognate disciplines such as crop improvement, crop production and crop protection. With the impetus for clean cultivation, information provided in the book can motivate researchers in developing environment-friendly and non-chemical means of improving crop production and activate the innate ability of the plant to enhance their field performance.

There is a strong need to focus on the implications that crop physiology research can contribute in ensuring food and nutritional security by way of understanding the mechanisms of yield formation, stress resilience and crop quality. This will be made possible by scientifically sound translational research on appropriate platforms. In this regard, it will be rewarding that novel insights on plant growth and developmental responses to diverse natural environmental cues be unraveled, which would pave way for genetic manipulations to enhance productivity.

Dr. R. Chandra Babu
Vice Chancellor
Kerala Agricultural University

Preface

The idea of the book is to help researchers and progressive farmers to put into practice the outcome of dedicated and conscientious research in some lesser applied and very practical fields of plant physiology. Information on the innate ability of the plant system to imbibe the different cues of nature and respond in ways, which have been considered allegoric have now been unraveled with advancement in technology. Many assumptions have been redefined with a strong scientific base.

This book contains seven chapters compiled by researchers working in the field of plantphysiology;viz.(1)"Paecapacityoftheplanttomoulditsphysiologicalprocesses and growth in accordance with the environment; (2) "Plant acoustic responses-concept and significance"- dealing with the response of plants to different types of sound, both natural and man-made; (3) "Spectral manipulation of plant responses"-light being a major constraint in productivity can have a major stake in improving yield potential especially in polyhouse cultivation; (4) "Geomagnetic responses in plants"- the influence of geomagnetism on plant life needs greater emphasis as this information may help to devise better management practices in crop production; (5) "Electricity from living plants-myth or reality"- highlighting clean energy as the need of the hour by utilizing green plants to produce electricity and to run electronic devices which needs further elucidation and encouragement; (6) "Plant architecture-evolution, diversity, regulation and scope"- manipulating plant growth for aesthetic value is a common practice but utilizing the same green architecture for utility is a more recent concept; (7) "Plant neurobiology-a paradigm shift in plant science"- a new branch of plant communication which is being recognized by plant scientists.

We hope that this book will immensely augment the knowledge gathering process of people who would like to have a better understanding of plant behavior. We pen our gratitude to all the contributors and well-wishers of this book. We also acknowledge the wholehearted support from New India Publishing Agency, New Delhi and their production team in this venture.

Authors

Contents

Abbreviations

ABA	Abscisic acid
AP	Action Potential
AP	Action Potential
ATP	Adenosine Triphosphate
CAM	Crassulacean Acid Metabolism
CAT	Catalase
DC	Direct Current
DNA	Deoxyribonucleic acid
EAB	Electron Active Bacteria
FAD	Flavin Adenine Dinucleotide
GMF	Geo Magnetic Field
GS-MS	Gas Chromatography-Mass Spectrometry
GUS-β	Glucuronidase
IAA	Indole-3-Acetic Acid
IC	Integrated Circuit
LAR	Leaf Area Ratio
LED	Light Emitting Diode
LED	Light emitting Diode
LMR	Leaf Mass Ratio
MF	Magnetic Field
MWNT	Multiwalled Carbon Nanotubes
NAA	Naphthalene Acetic Acid
NUE	Nutrient Use Efficiency
OAB	Oxygen Active Bacteria
PAFCT	Plant Acoustic Frequency Control Technology
PAR	Photosynthetically Active Radiation
PEDOT	Poly (3,4-ethylenedioxythiophene)
PEDOT:PSS	NFC-PEDOT:PSS combined with nanofibrillar cellulose
PEDOT:PSS	PEDOT doped with polystyrene sulfonate
PHY	B-Phytochrome B
PMFC	Plant-Microbial Fuel Cell
POD	Peroxidase
PR proteins	Pathogenesis Related proteins
PS II	Photosystem II
RLR	Root Length Ratio
RNA	Ribonucleic acid

ROS	Reactive Oxygen Species
SAA	Systemic Acquired Acclimation
SAM	Shoot apical meristem
SAR	Systemic Acquired Resistance
SLA	Specific Leaf Area
SMF	Static Magnetic Field
SOD	Superoxide Dismutase
SP	System Potential
SWP	Slow Wave Potential
TrpH	Tryptophan Reduced
UV	Ultraviolet
UV-C	Ultraviolet-C
VOCs	Volatile Organic Compounds
VOCs	Volatile Organic Compounds
WCMF	Weak Combined Magnetic Field
WP	Wound Potential

1

Phenotypic Plasticity Concepts and Recent Advances

Girija T **and** ***Parvathi M S***

Department of Plant Physiology, College of Horticulture
Kerala Agricultural University, Thrissur, Kerala

Introduction

Plasticity is a unique attribute by which the otherwise sessile plant species are capable of adjusting to environmental vagaries by changing their phenotypic and morphological characters. Any plant population will be able to survive only if it can respond to an extremely variable environment by becoming more plastic and more genetically variable. To achieve this, plants should be able to suitably change their morphology anatomy and physiology or any one of these parameters based on environmental changes so as to improve their adaptability and survival in the altered environment. When such changes are genetically heritable, then they lead to species evolution.

Plasticity helps plants to adapt to new environmental conditions, after migration to new geographical areas, by genetic assimilation of traits that can help to improve their performance under a new set of conditions. Understanding plasticity is important for predicting and managing the effect of climate change on native species as well as crop plants. Response of plants to modified environmental conditions is critical for their persistence.

Phenotypic plasticity has been defined as the ability of an individual genotype to produce different phenotypes when exposed to different environmental conditions (Pigliucci *et al.*, 2006). This includes the ability of a plant to modify its development in response to environmental cues and also its ability to bring about changes in its metabolism which increases the chances of survival of an organism under a specific condition.

Variation of a species under different environmental conditions has been referred to as individual variability by Darwin, 1859. The term phenotypic polymorphism was coined by Mayr (1963) which helps to explain environmentally induced phenotypic variations and distinguish it from genetic polymorphism.

Types of plasticity

Phenotypic plasticity is seen at every stage of plant growth and development. During the initial stages of growth, we observe more of adaptive responses that decides the plant architecture while during developmental stages, signal responses are more pronounced. According to (Nicotra *et al.,* 2010) phenotypic plasticity can be either long term or short-term. Long term plasticity is genetically controlled and heritable, which leads to evolution. Short term plasticity may be due to abiotic stress, herbivory or competition.

Physiological, morphological and anatomical plasticity can have different roles in enhancing plant adaptations.

Physiological plasticity ensures adjustment of gas exchange, light utilization and metabolic regulations that affect growth and development. This in turn enhances the capacity of species to colonize a new area and also acclimatize to adverse environments (Niinemets and Valladares, 2004; Zunzunegui *et al.,* 2009).

Morphological and anatomical plasticity seem to have only secondary roles; they bring about modifications in resource acquisition, allocation and structural characters which improve plant adaptations. This also depends on age, placement of plant parts and physiological activity level which changes with time. Competitiveness of a new species in an environment depends on resource acquisition which alters with development and heritability of these traits (Harper, 1985).

Since plasticity is an adaptive mechanism that allows a plant to optimally respond to environmental heterogeneity, the development of phenotypic polymorphism may be either adaptive, non-adaptive (maladaptive) or neutral.

Adaptive plasticity: This is an advantage as it allows a genotype to have broader tolerance to environmental conditions and has higher fitness across multiple environments (Bradshaw, 1965; Baker, 1974; Pigliucci, 2001, 2005).

Non-adaptive plasticity: Many of the environmentally induced variations are non-adaptive. This occurs when compared to the ancestral phenotype, the fitness of the new environmentally induced phenotype has a reduced average fitness value, which is much lower or is further away from the adaptive peak in the new environment. This type of non-adaptive plasticity can itself lead to a cryptic evolution and produce a different population in a stressful environment. For example, under moisture stress conditions, plants may fail to grow to an optional height and produce few seeds (Kleunen and Fisher 2005; Grether, 2005). However, among these by chance a small number of genotypes may exhibit a beneficial plastic response to allow a few individuals to persist long enough to survive and reproduce in the new environment and pass on the new material or epigenetic effect leading to adaptive evolution. Hence, the non-adaptive character by itself can become an adaptive mechanism in the altered environment.

When the plant is unable to compensate for environmental stress and it leads to deleterious influence on the plant it is known as *injurious plasticity*. For example, in an arid environment, if the plant cannot maintain high water potential or if a plant produces less extensive rhizome systems in a more compact soil, in both cases growth is affected (Schmid and Bazzaz, 1990).

However, when there is no response of a species to environment vagaries, with no effect on fitness, then it is known as *neutral plasticity* (Alpert and Simms 2002; Callahan *et al.,* 2005; Ghalambor *et al.,* 2007).

Factors influencing phenotypic plasticity

A number of heterogeneous environmental factors can contribute to developmental changes in plants. When these changes occur over days, weeks, or months those plants that can alter certain features better under the new conditions is more likely to survive and reproduce.

Light

One of the expressions of plant phenotypic plasticity is the modification of leaf traits in different light environments. The most important leaf trait affected is specific leaf area (SLA-the ratio between leaf area and leaf dry weight; Gratani, 2014). A strong correlation exists between leaf thickness and light saturation rate of photosynthesis per unit leaf area (Dorn *et al.,* 2000). Sun leaves have higher photosynthetic rate compared to shade leaves which is based on leaf lamina thickness per unit leaf area (McClendon and McMillen, 1982). Another important trait influenced by light is the anthocyanin pigmentation. Anthocyanins are produced in leaves in response to excess light, temperature and osmotic extremes, and serve as a reversible plastic mechanism for the protection of photosynthetic machinery. Growth of internodes and leaf angles are affected by both quantity and quality of light (Franklin and Whitelam, 2005)

Temperature

Plant species have an optimum range of temperature requirement for growth and development. When temperatures are higher or lower than normal, it may contribute to abnormal growth such as elongation of petiole and increased water loss leading to reduction in yield (Crawford *et al.,* 2012). Temperature cues are important for floral evocation; hence flowering time is an important character that has shown to be under both environmental and genetic control. Changes in temperature triggers a chain of events which induces floral initiation and flowering, provided the plant has completed its juvenile stage. Due to climate change, such cues may not be reliable if they occur at the wrong time with respect to the lifecycle and ecology of a species. Induction of such environmental cues or signals which can elicit a response scheme might contribute to maladaptive plastic changes in plants.

Water

In addition to water, various other factors such as soil temperature, nutrients and pH affect root architecture of plants. Adaptation of roots to various environmental factors is very important in crop growth and yield. Morphological plasticity of root characters such as root tip diameter, gravitropism and rhizo-sheaths allow the plants to adapt to new situations which may also be useful for improving water use efficiency in crop species. Such plastic responses have been observed in *Mesembryanthemum crystallinum,* a native of Africa, Western Asia and Europe. The plant normally has a prostrate succulent nature and is covered with large glistening bladder cells or water vesicles. It is also commonly known as ice plant or crystalline ice plant. Bladder cells are enlarged epidermal cells. The main function of 'bladder' is to reserve water. Under normal condition, the plant has C_3 metabolism but when exposed to water or salt stress, it shifts from C_3 metabolism to Crassulacean Acid Metabolism (CAM). Under CAM condition, they keep their stomata closed during the day and open at night which helps to conserve water. CAM pathway in the plant was seen to be rapidly reversible which was influenced by temperature, high light (Haag-Kerwer *et al.*, 1992) and drought stress (Borland *et al.*, 1992; de Mattos and Lüttge, 2001). Light-dependent stomatal opening and operation of xanthophyll cycle were studied in guard cells isolated from ice plants shifting from C_3 metabolism to CAM (Tallman *et al.*, 1997, Dodd *et al.*, 2002). It was seen that stomatal response of ice plants in the C3 mode depends solely on the guard cell response to blue light.

Soil type

Studies indicate that soil type and altitude affect the above ground resource accumulation and flowering phenology of plants (Haukka *et al.*, 2013). Such plastic responses are commonly observed in invasive weed species. Table 1 indicates changes in the duration of phenophases and the fecundity of the grass weed *Isachne miliacea* grown in different soil types of Kerala. Among the morphological characters, it was seen that leaf area, internodal length and seedling height are major attributes influenced by soil type (Fig. 1; Suada, 2015). This result clearly shows that the same species can express differently depending on the soil condition.

Table 1. Phenophases (days) of *I. miliacea* grown in different soil types

Phenophases	Onattukara	Kole	Kuttanad	Palakkad
Days to Seed germination	8–10	8-12	5–8	12–15
Days to Tillering	15–19	18–21	13–17	20–25
Days to Flowering	36–52	38–55	28–60	41–57
Days to Seed formation	40–62	42–65	33–68	50–62
Days to Seed maturation	54–70	55–68	40–74	56–70

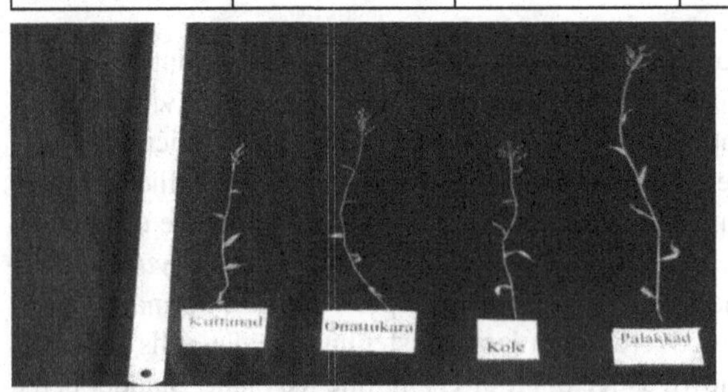

Fig. 1: Branches of *I. miliacea* grown in different soil types showing different internodal lengths

Competition

Neighboring plants seem to impart plasticity in plant species. Species differ in resource uptake abilities and plant communities are normally in a continuous state of war with each other for various growth factors. Competition among aerial and underground tissues for nutrients, light, water, space, CO_2 and O_2 can affect growth and yield of crop plants. Success in competition however, is largely dependent on the capacity of species for resource capture. Competitive ability of plants depends upon their aggressiveness in acquisition of growth factors and utilizing them in rapid tissue build-up for growth. Plasticity helps the plant to navigate away from the site of competition and help in resource capture. Root plasticity in response to neighbours is striking in natural communities. Competition induced plastic responses in root allocation and architecture is an important aspect of survival.

Herbivory

Herbivory activates a number of elicitors in plants which in turn activate numerous genes that produce PR proteins (pathogenesis-related). Some of these are antimicrobial and attack bacterial cell walls. Others spread "news" of the infection to nearby cells and this will activate the production of volatile organic substances which will act as signals to neighboring plants to initiate a number of physiological and biochemical changes in the plant system and contribute to hypersensitive reaction. Hence, plasticity is a major factor that decides tolerance or resistance of a species to herbivory.

Plasticity induced by interaction

Parasitic plants induce phenotypic plasticity in the host plant. It is seen that susceptible plants like tomato, produce volatiles such as terpenoids, α-pinene, β-myrcene, and β-phellandrene, which act as chemical cues to serve as chemo-attractants for parasites such as *Cuscuta.* Once they find a suitable host, the first physical contact initiates an attachment phase, in which the parasitic epidermal and parenchymal cells begin to differentiate into a secondary meristem and develop pre-haustoria, also known as adhesive disk (Dörr, 1968; Heidejorgensen, 1991). After attachment to the host, the pre-haustoria develops into parasitic haustoria that penetrate the host stem through a fissure. This is by exerting mechanical pressure (Dawson et al., 1994) and also by biochemical degradation of host cell walls (Vaughn, 2003). When they come in contact with sieve cells, these parasitic cells take up the function of both sieve elements and transfer cells and from there they form a connection to the xylem, after which the parasitic and host cells of the xylem parenchyma will synchronize their development and fuse together to form a continuous xylem tube from the host to the parasite (Dörr, 1972). Host plant then supplies the parasitic plant with water, nutrients, and carbohydrates through these functional connections to the xylem and phloem (Jeschke *et al.,* 1994; Jeschke *et al.,* 1997; Hibberd *et al.,* 1999; Hibberd and Jeschke, 2001).

However, in resistant tomato varieties, it is seen that the searching hypha of *C. reflexa* is blocked by a haustorial searching hypha. When the parasitic haustoria is formed after about 3-5 days at the end of the attachment phase, the epidermal host cells at the contact sites will elongate and burst. Moreover, in resistant tomato varieties,

C. reflexa is found to induce a defense program, in which tomato cells at the infection site secrete soluble phenyl propanoids and show an increased accumulation and activity of peroxidases. These enzymes bring about changes in composition of the cell wall by influencing the proteins, pectins or cellulose fiber content of the cell wall. When the cell wall is thus modified it prevents *C. reflexa* from penetrating the tomato plant.

Molecular basis of plastic responses in key traits

Expression of plasticity in a given trait is possible only if it is mediated at the molecular level. Environmental signals provide the impetus for developmental transformation this is by activating signaling pathways that sense abiotic cues such as light, nitrogen and drought, as well as biotic signals such as *Nod* factors that cause nodulation in legume.

Assessing plastic responses in plants

Plasticity is a major source of phenotypic variation which can influence selection. Adaptive diversity (Sultan, 2004) facilitates short term adaptation to environmental changes which can have practical applications in agriculture.

A few easily quantifiable plant functional traits will help to describe the ecology of a species. A number of such plant traits/characters have been identified to measure phenotypic plasticity in plants; their applicability may not be universal but depending on the plant species, these indices may be selected for comparison (Weigelt and Jolliffee, 2003, Sanchez-Gomez *et al.*, 2006; Gratani *et al.*, 2014). Some such characters have been listed below:

1. Plant height
2. Number of shoots
3. Internodal length
4. Partitioning
5. Leaf Area Ratio (LAR)
6. Root Length Ratio (RLR)
7. Nutrient Use Efficiency (NUE)
8. Specific Leaf Area (SLA)
9. Leaf Mass Ratio (LMR)
10. Leaf angle
11. Total leaf thickness
12. Leaf density
13. Stomatal density

Advantages and disadvantages of plant plasticity

The capability of plants to change their structure and function in the face of environmental changes can contribute to "selection" in the case of more plastic varieties of crops. These varieties may not necessarily be the most productive, nor have the most easily-predictable productivity. Although phenotypic plasticity can be very advantageous for plants, differences among species and populations in their plasticity may reflect differential selective pressures on plants. In nature, plants are exposed to several abiotic factors simultaneously. Ability of the plants to tolerate such complexes or more than one abiotic stress in general, are very scarce (Niinemets and Valladares, 2006). A combination of two or more stresses like low light and drought, in particular, has been suggested as a very strong ecological filter. Similarly, drought and heat stresses which have very strong implications on crop growth and productivity independently, coexist in field conditions. There is an emerging relevance to understand the plastic mechanisms of plants/crops to combined drought and heat stresses (Parvathi *et al.*, 2020).

Conclusion

Traditionally, agricultural plant breeders have viewed plasticity as an unwanted complication (Johnson and Frey, 1967), but perspectives on that are changing (Bradshaw, 2006; Chapman, 2008; Forde 2009; Sadras *et al.*, 2009). Knowledge about the genetic mechanisms underlying phenotypic plasticity (Schlichting and Smith 2002; Reymond *et al.*, 2003; Forde, 2009), will help plant breeders to improve performance over a broad range of conditions and also act as a buffer against extinction, improve ecological adaptation and help evolution. This can influence plant's chances of survival, reproduction and colonization and can make important contributions to improving yield stability in agriculture. Non-adaptive plasticity increases the strength of selection.

References

Alpert, P. and Simms, E.L. 2002. The relative advantages of plasticity and fixity in different environments: when is it good for a plant to adjust? *Evol. Ecol.* 16: 285–297.

Baker, H.G. 1974.The evolution of weeds. *Ann. Rev. Ecol. Syst.* 5: 1–24.

Borland, A. M., Griffiths, H., Maxwell, C., Broadmeadow, M.S.J., Griffiths, N.M., and Barnes, J.D. 1992. On the ecophysiology of the Clusiaceae in Trinidad- expression of CAM in *Clusia minor* L. during the transition from wet to dry season and characterization of three endemic species. *New Phytol.* 122:349–357.

Bradshaw, A. D. 1965. Evolutionary significance of phenotypic plasticity in plants," *Adv. Genet.* 13:115–155.

Bradshaw, A. D. 2006. Unraveling phenotypic plasticity – why should we bother? *New Phytol.* 170:644–648.

Callahan, H.S., Dhanoolal, N. and Ungerer, M.C. 2005. Plasticity genes and plasticity costs: a new approach using an Arabidopsis recombinant inbred population. *New Phytol.* 166: 129–139.

Chapman, S. C. 2008. Use of crop models to understand genotype by environment interactions for drought in real-world and simulated plant breeding trials. *Euphytica* 161: 195–208.

Crawford, A. J., McLachlan, D. H., Hetherington, A. M., and Franklin, K.A. 2012. High temperature exposure increases plant cooling capacity. *Curr. Biol.* 22(10): 396–397.

Darwin, C.1859. *On the Origin of Species by Means of Natural Selection, or the Preservation of Favoured Races in the Struggle for Life.* John Murray, Albemarle Street, London. 44p.

Dawson, J. H., Musselman, L. J., Wolswinkel, P., and Dörr, I. 1994. Biology and control of *Cuscuta, Musselman, Rev. Weed Sci.* 6:265–317.

de Mattos, E.A. and Lüttge, U. 2001. Chlorophyll fluorescence and organic acid oscillations during the transition from CAM to C_3-photosynthesis in *Clusia minor* L. (Clusiaceae). *An.* Bot. 88:457–463.

Dodd, A.N., Borland, A.M., Haslam, R.P., Griffiths, H., and Maxwell, K. 2002. Crassulacean acid metabolism: plastic, fantastic. *J. Exp. Bot.* 53:569–580.

Dorn, L.A., Pyle, E. H., and Schmitt, J. 2000. Plasticity to light cues and resources in *Arabidopsis thaliana*: testing for adaptive value and costs. *Evol.* 54(6): 1982–1994.

Dorr, I. 1968. Localization of cell contacts between *Cuscuta odorata* and different higher hostplants. *Protoplasma* 65: 435–448.

Dorr, I. 1972. Contact of *Cuscuta*-Hyphae with sieve tubes of its host plants. *Protoplasma* 75: 167–187.

Forde, B.G. 2009. Is it good noise? The role of developmental instability in the shaping of a root system. *J. Exp. Bot.* 60: 3989–4002.

Franklin, K.A. and Whitelam, G.C. 2005. Phytochromes and shade-avoidance responses in plants. *Ann. Bot.* 96:169–175.

Ghalambor, C.K., McKay, J.K., Carroll, S.P., and Reznick, D.N. 2007. Adaptive versus non-adaptive phenotypic plasticity and the potential for contemporary adaptation in new environments. *Funct. Ecol.* 21: 394–407.

Gratani, L. 2014. Plant phenotypic plasticity in response to environmental factors [on-line]. Available: https://doi.org/10.1155/2014/208747 [05 Nov. 2019].

Gratani, L., Crescente, M. F., D'amato, V., Ricotta, C., Frattaroli, A.R., and Puglielli, G. 2014. Leaf traits variation in *Sesleria nitida* growing at different altitudes in the Central Apennines. *Photosynthetica* 52(3):386-396.

Grether, G.F. 2005. Environmental change, phenotypic plasticity, and genetic compensation. *Am. Nat.* 166:115–123.

Haag-Kerwer, A., Franco, A.C., and Lüttge, U. 1992. The effects of temperature and light on gas exchange and acid accumulation in the C3–CAM plant *Clusia minor* L. *J. Exp. Bot.* 43:345–352.

Harper, J. L. 1985. Modules, branches and the capture of resources. In: Jackson, J.B.C., Buss, L.W., and Cook, R.E. (eds), *Population Biology of Clonal Organisms*. Yale University Press, New Haven, Connecticut, USA, pp.1-33.

Haukka, A.K., Dreyer, L.L., and Esler, K.J. 2013. Effect of soil type and climatic conditions on the growth and flowering phenology of three Oxalis species in the Western Cape, South Africa. *S. Afr. J. Bot.* 88:152-163.

Heidejorgensen, H. S. 1991. Anatomy and ultrastructure of the haustorium of *Cassytha-Pubescens* R. Br. I. The adhesive Disk. *Bot. Gaz.* 152: 321–334.

Hibberd, J. M. and Jeschke, W. D. 2001. Solute flux into parasitic plants. *J. Exp. Bot.* 52:2043–2049.

Hibberd, J. M., Quick, W. P., Press, M. C., Scholes, J. D., and Jeschke, W. D. 1999. Solute fluxes from tobacco to the parasitic angiosperm *Orobanche cernua* and the influence of infection on host carbon and nitrogen relations. *Plant Cell Environ.* 22: 937–947.

Jeschke, W. D., Baig, A., and Hilpert, A. 1997. Sink-stimulated photosynthesis, increased transpiration and increased demand-dependent stimulation of nitrate uptake: nitrogen and carbon relations in the parasitic association *Cuscuta reflexa* - *Coleus blumei*. *J. Exp. Bot.* 48 915–925.

Jeschke, W. D., Baumel, P., Rath, N., Czygan, F. C., and Proksch, P. 1994. Modeling of the flows and partitioning of carbon and nitrogen in the holoparasite *Cuscuta reflexa* Roxb. and its host *Lupinus albus* L.:II. Flows between host and parasite and within the parasitized host. *J. Exp. Bot.* 45: 801–812.

Johnson, G.R. and Frey, K.J. 1967. Heritabilities of quantitative attributes of oats (*Avena* sp.) at varying levels of environmental stress. *Crop Sci.* 7, 43-46.

Kleunen, M. V. and Fischer, M. 2005. Constraints on the evolution of adaptive phenotypic plasticity in plants. *New Phytol.* 166: 49–60.

Mayr, E. 1963. *Animal species and Evolution.* Harvard University Press, London, 797p.

McClendon, H. and McMillen, G.G. 1982. The control of leaf morphology and the tolerance of shade by woody plants. *Botanical Gaz.* 143(1):79–83.

Nicotra, A.B. and Davidson, A. 2010. Adaptive phenotypic plasticity and plant water use. *Funct. Plant Biol.* 37:117-127.

Niinemets, U. and Valladares, F. 2004. Photosynthetic acclimation to simultaneous and interacting environmental stresses along natural light gradients: optimality and constraints. *Plant Biol.* 6(3): 254–268.

Niinemets, Ü. and Valladares, F. 2006. Tolerance to shade, drought, and water logging of temperate Northern Hemisphere trees and shrubs. *Ecol. Monogr.* 76:521–547.

Parvathi, M.S., Dhanyalakshmi, K.H., and Nataraja, K.N. 2020. Molecular mechanisms associated with drought and heat tolerance in plants and options for crop improvement for combined stress tolerance. In: Hasanuzzaman, M. (ed.), *Agronomic Crops*, Springer, Singapore, pp. 481-502.

Pigliucci, M. 2001. *Phenotypic Plasticity: Beyond Nature and Nurture*. John Hopkins University Press, Baltimore. 328p.

Pigliucci, M. 2005. Evolution of phenotypic plasticity: where are we going now? *Trends Ecol. Evol.* 20(9):481–486.

Pigliucci, M., Murren, C. J., and Schlichting, C. D.2006. Phenotypic plasticity and evolution by genetic assimilation. *J. Exp. Biol.* 209:2362–2367.

Reymond, M., Muller, B., Leonardi, A., Charcosset, A., and Tardieu, F. 2003. Combining quantitative trait loci analysis and an ecophysiological model to analyze the genetic variability of the responses of maize leaf growth to temperature and water deficit. *Plant Physiol.* 131: 664-675.

Sadras, V.O., Reynolds, M.P., de la Vega, A.J., Petrie, P.R., and Robinson, R. 2009. Phenotypic plasticity of yield and phenology in wheat, sunflower and grapevine. *Field Crops Res.*110: 242–250.

Sanchez-Gomez, D., Valladares, F., and Zavala, M.A. 2006. Functional traits and plasticity in response to light in seedlings of four Iberian forest tree species. *Tree Physiol.* 26(11):1425–1433.

Schlichting, C. D. and Smith, H. 2002. Phenotypic plasticity: linking molecular mechanisms with evolutionary outcomes," *Evol. Ecol.* 16(3):189-211.

Schmid, B. and Bazzaz, F.A. 1990. Plasticity in plant size and architecture in rhizome-derived solidago and aster. *Ecol.* 71(2):523-535.

Suada, A. P. 2015. Growth and physiology of *Isachne miliacea* Roth. in different soil types and its sensitivity to common herbicides. M.Sc. Thesis, Kerala Agricultural University, Thrissur.

Sultan, S. E. 2004. Promising directions in plant phenotypic plasticity," *Perspect. Plant Ecol., Evol. Syst.* 6(4):227–233.

Tallman, G., Zhu, J., Mawson, B.T., Amodeo, G., Nouhi, Z., Levy, K., and Zeiger, E.1997. Induction of CAM in *Mesembryanthemum crystallinum* abolishes the stomatal response to blue light and light-dependent zeaxanthin formation in guard cell chloroplasts. *Plant Cell Physiol.* 38 (3): 236–42.

Vaughn, K. C. 2003. Dodder hyphae invade the host: a structural and immunocytochemical characterization. *Protoplasma* 220: 189–200.

Weigelt and Jolliffe, P. 2003. Indices of plant competition. *J. Ecol.* 91(5):707–720.

Zunzunegui, M., Ain-Lhout, F., Barradas, M.C.D., Álvarez-Cansino, L., Esquivias, M.P., and García Novo, F. 2009. Physiological, morphological and allocation plasticity of a semi-deciduous shrub," *Acta Oecol.* 35(3):370–379.

2

Plant Acoustic Responses Concept and Significance

Gayathri Rajasekharan **and** ***Nandini K.***
Department of Plant Physiology, College of Horticulture
Kerala Agricultural University, Thrissur, Kerala

Introduction

Plants possess complex, systematic and specific mechanisms to cope up with abiotic as well as biotic stresses in the existing environment. 'Do plants respond to sound? How can plants perceive sound without a hearing organ?'- are interesting queries among the scientific community. Responses of plants to different factors such as temperature, moisture, light, wind, pests and microbes are well understood. However, information on the effects of audible sound frequencies on plants are still lacking. Recent evidences suggest that audible sound stimulation can contribute to better plant growth and improved quality of plant produce. It also represents a new trigger for plant protection. Sound is an acoustic energy in the form of a mechanical wave which transmits through gases, liquids and solids. Acoustic spectrum consists of three regions, infra sound (less than 20 Hz), audible region (20-20,000 Hz) and ultrasound (greater than 20,000 Hz). Infra sound and ultrasound are used in clinical diagnosis and therapeutics.

Research on plant responses towards different stimuli started during 1880s by Charles Darwin. In 1848, Dr. Gustav Theodor Fechner, a German experimental psychologist, suggested that plants are capable of feeling emotions and that one could promote healthy growth of plants with talk, attention, attitude and affection. Later, J. C. Bose (1926), famous Indian botanist and physicist, made several studies on the effect of sound vibrations on plants. According to him, a plant treated with care and affection gives out a different vibration than a plant subjected to torture. In addition, Bose found that plants grew more quickly in pleasant music and more slowly in loud noise or harsh sounds. He also noticed that plants can "feel pain and understand affection" based on variation of cell membrane potential under different sound frequencies. Retallack (1973) also described several experiments involving plants and music.

Sound acts as an alternate stress on plants and can either promote or restrict growth depending on the frequency of sound used. Acoustic frequency alters thermodynamic behaviour, secondary structure of protein and fluidity of cell membranes (Meng *et al.*, 2012). Moreover, decrease in phase transition temperature as well as enhancement of fluidity of cell wall and membrane results in faster division of cells. In addition, plants can produce sound waves at relatively low frequencies of 50-120 Hz spontaneously

(Hassanien *et al.*, 2014). Plants also can absorb and resonate to specific external sound frequencies.

Table 1. Important research outcomes of studies on the effect of sound on plants

S. No.	Responses	References
1.	Plants are capable of feeling emotions	Fechner, 1848
2.	Plants can feel pain and understand affection	Bose, 1926
3.	Classical music cause two fold increase in growth than plants not exposed to music	Singh, 1962
4.	Increases yield in rice Increases yield in wheat	Subramanian *et al.*, 1969 Weinberger and Measures, 1979
5.	Increases fluidity of cell membrane Decreases the phase transition temperature	Xiujuan *et al.*, 2003
6.	Increases plasmalemma H^+ ATPase activity	Yi *et al.*, 2003
7.	Increases content of growth hormone	Meng *et al.*, 2012
8.	Enhances seed germination Stimulates stomatal opening Stress-induced genes are switched on	Hassanien *et al.*, 2014

Plant Acoustic Frequency Control Technology (PAFCT)

Plant Acoustic Frequency Control Technology (PAFCT) or sonic growth-promotion technology, is a new technique which uses an acoustic frequency generator to produce appropriate acoustic waves that matches with the sound frequency of specific plants. It aims to expose plants to sound waves in special frequency which is in harmony with the plant's meridian system so that it can increase plant growth, development and enhance resistance (Hou *et al.*, 1994). According to Xiujuan *et al.*, 2003, response of plant depends on the intensity, frequency, time and period of exposure to sound waves.

Plant cell membranes are equipped with large number of mechano-sensitive channels which are responsive to mechanical vibrations (Haswell *et al.*, 2011). Sound vibrations (SV's) can cause an alteration in biological membranes that could possibly evoke a signalling cascade through activation of these channels. At the cellular level, SVs can change the secondary structure of plasma membrane proteins, affect microfilament rearrangements, produce Ca^{2+} signatures, cause increase in protein kinases, protective enzymes, peroxidases, antioxidant enzymes, amylase, H^+-ATPase/K^+ channel activities, and enhance levels of polyamines, soluble sugars and auxin (Mishra *et al.*, 2016).

Different plant species have various responses to sound stimulation at different growth stages. Acoustic biology has become increasingly popular and more attention is accorded to the effects of environmental stresses on the growth and development of plants.

Types of acoustic responses

Sound has growth promoting as well as inhibiting effect on plants. Some reports indicate that classical music can favour growth (Ekici *et al.*, 2007); classical music has a gentle vibration, and it's easy on plants. Violin music in particular is known to significantly increase plant growth. Sound can also have detrimental effects on plant growth. Music containing hard-core vibrations could be devastating to plants. Even played at low volumes, heavy metal music can be very damaging to a sensitive plant (Chivukula and Ramaswamy, 2014).

Effects of sound on plants

Sound can influence different aspects of plant growth; based on available information they can be classified as

1. Morphological changes
2. Physiological changes
3. Biochemical changes
4. Changes in yield and quality

Morphological changes induced by sound waves

Seed germination, plant growth and development including root elongation and plant height are some of the growth parameters influenced by different frequencies of sound.

Effect of sound on seed germination

Bochu and co-workers in 2003 conducted experiments on sound stimulation using paddy seeds. He had two groups of five teams, which contained 50 drops of seeds in each team. A team without stimulation was maintained as control team in both groups. In group 1, seeds were stimulated under 400 Hz with different intensities (96, 101, 106 and 111 dB). In group 2, seeds were stimulated under 106 dB with different frequencies (200, 400 Hz, 1 and 4 kHz). Each group was stimulated twice a day (30 min at one time) for 2 days and then put into cultivation-tanks with steady-temperature and lighting, at 25.8 °C for 5 days.

Under the sound wave exposures of 200 and 400 Hz, the germination index, height of stem, fresh weight and number of roots showed an increase. When the frequency reached 1000 Hz, there was a swing in the opposite direction. With a stimulation frequency of 4000 Hz, plants were damaged. Similarly, when the sound intensity was less than 400 Hz, it showed beneficial effects and a reverse effect was observed when the intensity was increased beyond the normal limit. This experiment showed that 400 Hz and 106 dB is the 'best frequency and intensity' combination to activate seed germination and growth. It was also inferred that sound stimulation had a positive function only when the frequency of stress coincided with the biological frequency of the cell.

The study brought out the beneficial effects of sound which suggested that suitable sound stimulation may have contributed to absorption of nutrients and transformation of energy, resulting in accelerated germination and growth of seeds. Sound possibly caused improvement of protein content in germinating seeds. This may be due to three reasons:

1. Sound field should have influenced the cell cycle of seed cell and speeded up its reproduction rate.
2. It should have played a part in the transfer of energy into the cell and in nutrient uptake.
3. Sound could have affected the membrane biophysics to cause changes of biological function of membrane and enhance cell metabolism.

Sound is known to affect various metabolic activities involved in seed germination. Experiments conducted to compare the effects of music, noise and healing energy on seed germination revealed that musical sound significantly enhanced seed germination in okra and Zucchini seeds (Creath and Schwartz, 2004). Chuanren *et al.,* (2004) observed highest seed germination with least germination time when seeds of *Echinacea angustifolia*, a medicinal plant, were exposed to sound waves at 100 dB and 1000 Hz.

Sound stimulation increases the cell wall and membrane integrity, which facilitated cell division and growth (Zhao *et al.,* 2002). Sound stimulation also enhanced the fluidity of physical state of lipids in plasma membrane and favoured secondary structure of proteins in cell wall and plasmalemma. These structural changes of protein and membrane moieties supported the membrane affecting modulation of metabolic activity (Yi *et al.,* 2003).

Effect of sound on growth and development

Several studies have been undertaken to study and understand the influence of sound and music on plants and plant growth. Chivukula and Ramaswamy (2014) studied the effect of music on rose (*Rosa chinensis*) plants grown in separate pots. He divided the plants into five groups and each group was exposed to 4 different types of music which included rock music, Indian and western classical music and vedic chants; the control plants were kept in silence. Observations were recorded over a period of 60 days on shoot elongation, internodal elongation, number of flowers and flower diameter.

Plants exposed to vedic chants showed maximum shoot elongation, maximum number of flowers and highest diameter of flowers. Internodal elongation was highest in plants exposed to Indian classical music. Exposing plants to vedic chants or Indian classical music improved plant growth as compared to those kept in silence, western classical or rock music. Plant growth was observed to be optimum when they were exposed to pure tones in which the wavelength of sound coincides with the average of major leaf dimensions. Hence, it may be presumed that plants too enjoy music and their preference of music may be linked to the wavelength of sound. Higher growth indicates, faster nutrient uptake, better assimilation and rapid cell division.

It was also observed that music containing hard-core vibrations could be harmful to plants. Music with high vibrations, such as rock music does not allow plants to grow at their normal pace. In the above experiment, it was ensured that the volume of the music was equal for all the music types with which they were able to rule out the difference in volume, so the source of variation was the other factor that contributed to difference in growth of plants.

The increased rate of growth in terms of more flowers, leaves, buds *etc.* suggests that specific audible frequencies including music can benefit agriculture by increasing the productivity as reported by Chowdhary *et al.*, (2014). Playing ancient traditional Indian chants makes a remarkable change in the growth of plants. Variable growth of shoots and leaves were observed with the highest shoot length measured from the plants provided with Sanskrit *shlokas* as reported by Ankur *et al.*, 2016.

Effect of sound on root growth

Root elongation is related to cell metabolism and a positive relationship between root growth and different types of music have been reported. Seregin and Ivanor (2001) observed a positive correlation between root elongation, mitotic division and rhythmic classical music.

Stems from chrysanthemum seedlings were inoculated in conical flasks with 20 ml MS solid medium (supplemented with 0.001 mg/L IAA) and cultured in an illumination incubator at 26°C. Inoculated stems were stimulated by sound waves with certain intensity (100 dB) and frequency (1000 Hz) for 3, 6, 9, 12 and 15 days, respectively; each day for 60 min. The control group was placed in the same environment as that of the stressed groups. After 9 days of stimulation of chrysanthemum plants, observations were taken after one month. Root fresh weight, root length and activity of root increased greatly under sound wave stimulation (Yi *et al.*, 2003). Number of roots had no apparent difference, but roots subjected to stimulation were thicker than the control roots. Results indicate that sound waves with 1000 Hz and 100 dB can increase the content of soluble proteins in chrysanthemum roots, which is advantageous for plant cell growth and splitting. The study shows how sound stimulation can promote growth and development of plants when strength and frequency of sound wave are correct. Root activity is an important marker reflecting root metabolism. Recent evidences illustrated that young root tips of maize show bending towards a continuous sound source and high bending percentage was measured between 0.2 and 0.3 kHz (Hassanien *et al.*, 2014).

Ekici *et al.*, (2007) investigated the effects of two types of music groups: group 1 indicated strong, complex, rhythmic accent classical music whereas group 2 was rhythmic, dynamically changing lyrics of classical music. Both music types had positive effects on root growth and mitotic divisions in onion root tip cells. Rhythmic dynamically changing lyrics gave better results with regards to the length of adventitious roots, while dynamically changing lyrics improved growth and mitotic index value as compared to control.

Physiological changes induced by sound waves

Sound has a significant effect in evoking changes in chlorophyll content and photosynthesis. To find out the effect of sound on chlorophyll content, strawberry seedlings at four leaf stage were subjected to Plant Acoustic Frequency Control Technology (PAFCT) at an intensity of 100 dB and a frequency of 40-2000 Hz respectively 3 hours every day (Meng *et al.,* 2012), lasting from 7am to 10am based on previous research results. After treatment for 42 days (21 times), all the plants were tested for chlorophyll content, net photosynthetic rate and photochemical efficiency of PS II. The chlorophyll content of strawberry increased by acoustic frequency treatment. There was no significant difference in chlorophyll content between cytokinin treatment and sound treatment when the plants were tested 30 days after transplanting. However, after 40 days and 20 times of PAFCT, there was a significant difference, wherein sound treated plants had higher chlorophyll than cytokinin treated plants. This fact revealed that it took 40 days for sound treatment on strawberries, to show significant variation in chlorophyll content.

In the same experiment, significant differences in photosynthetic rate was observed between control and treatment groups, 45 days after transplanting. The reasons for the improvement in photosynthetic capacity of strawberry with PAFCT may be attributed to two main aspects. The first includes factors involving absorption, transformation and transfer of light energy, such as degradation of chloroplast pigment, increase in electron transfer and photochemical efficiency caused by increase in light-harvesting pigment complexes. The second aspect comprises of carbon assimilation, including stomatal conductance, Rubisco activity and attenuated mesophyll resistance which results from elevated relevant enzyme activity in Calvin cycle. The study also revealed that acoustic frequency treatment could improve the activity of photosystems in the reaction centre and enhance electron transport and photochemical efficiency of PS II. PAFCT test in mustard was able to increase average plant height by 10.4%, leaf area by 28.4%, total chlorophyll content by 27.7%, nitrogen content by 25.436% and potassium content by 34.42% as compared to control (Zakariya *et al.,* 2017).

Biochemical changes induced by sound waves

Sound waves at specific frequencies and strength also has an effect on nucleic acid, protective enzyme activity, H^+ATPase activity and hormone content in plants. Among these, nucleic acid is an important biomolecule which can be inherited (Avery, 1940). To understand the effect of sound on the biological system, Xiujuan and co-workers in 2003 studied the effect of sound on the nucleic acid and soluble protein contents. This was to understand the actual influence of sound on gene expression. It is the level of DNA and RNA that brings about changes in physiological and metabolic processes of plants by controlling transcription, post-transcription, translation and post-translational modifications to stimulate a response in plants (Isono *et al.,* 1997).

Yi *et al.,* 2003, tried to study the effect of sound stimulation on nucleic acid content of chrysanthemum stems. He found that DNA content did not show much variation while RNA content showed significant variation; highest RNA content was observed when the period of stimulation was 9 days. There was variation in soluble protein level when the period of induction was changed. Highest value was observed when plants were exposed for 9 days. This clearly indicates that sound waves are capable of upregulating some stress induced genes and this in turn would have activated the level of transcription improving the content of RNA, thereby stimulating translation process leading to increased protein content. However, the mode of perception of sound waves by the cell and process of gene activation needs further elucidation. In addition to this, soluble sugars, proteins and amylase activity in chrysanthemum increased significantly with sound wave of intensity 100 dB and frequency 1000 Hz (Yi *et al.,* 2003).

Plant cells generate free radicals such as O_2^-, $OH^\cdot$ and $^1O_2$ in multiple organelles by different metabolic processes throughout their growth period. Free radicals have strong oxidative capacity and can destroy many functional molecules. Protective enzymes such as peroxidase (POD), catalase (CAT) and superoxide dismutase (SOD) help to scavenge these free radicals from plant system and maintain their content at a safe level in the cells. A study conducted by Xiujuan *et al.,* 2003, in chrysanthemum seedlings reveled that activities of CAT, POD and SOD increased at varying degrees when plants were subjected to 1.4 kHz, 0.095 kdB sound waves for different stimulation periods. Experimental data demonstrated that activities of SOD increased by 3 days while POD and CAT increased by 6 days of stimulation. Highest value was recorded for 9 days stimulation period and thereafter a decline in values was noticed, though higher than the control group. This experiment revealed that stimulation with sound waves with certain intensity and frequency can increase activities of enzymes in plants and promote plant growth and development.

Hormones have been suggested to play a prominent role in the control of callus growth. Bochu *et al.,* 2004 investigated the induction effect of sound waves on dynamic changes in endogenous levels of indole-3-acetic acid (IAA) and abscisic acid (ABA) *in vitro* during differentiation process of mature callus of chrysanthemum. Explants from tube culture of chrysanthemum were induced in cone-shaped flasks containing 30 ml Murashige and Skoog (MS) solid medium supplemented with 1.0 mg L^{-1} 2-naphthyl acetic acid (NAA) and 1.5 mg L^{-1} 6-benzyl amino purine (6-BA) in the treated group and control ones, respectively. Explants were kept in continuous cool-white fluorescent light, 1200 lux for 16 h per day at 24°C, and 70–80% relative humidity. Callus of the treated groups were exposed to optimal sound wave conditions of 1.4 kHz, 0.095 kdB based on previous results for half an hour each two times daily.

Groups treated by optimal sound wave (1.4 kHz, 0.095 kdB) had significantly higher IAA levels and lower ABA than that of control, increased level of IAA as well as decreased levels of ABA correlated with sound wave stimulus. Since high rate of IAA/ABA was favourable for development and differentiation of callus, they concluded that sound waves influence hormonal balance in plants which had a direct impact on callus growth. However, the molecular mechanism of sound wave perception and transduction of active response of hormonal regulation are not clear at present.

Effect of sound treatment under water stress conditions

Research on response of sound treated *Arabidopsis* adult plants to water stress revealed that survival percent of drought induced plants was higher by 24.5% in sound treated plants as compared to plants kept in silence, which was only 13.3% (Lopez-Ribera and Vicient, 2017). RNA-sequencing revealed significant upregulation of 81 genes involved in abiotic stress response (32 genes), pathogen response (31 genes), oxidation reduction process (11 genes), regulation of transcription (5 genes) and in protein phosphorylation-dephosphorylation.

Response of plants to nature's music

Effect of cuckoo song and cricket's chirps

In nature, plants are exposed to the music of insects and birds. Jun and Shi-ren (2011) made an attempt to understand the effect of insect-music mixed sound, cuckoo acoustic song and cricket acoustic song on the growth of cowpea seedlings. Since these sound waves differed in frequency, he also designed a single 400 Hz frequency sound wave and Fn and F5 sound waves composed of different frequencies and included them in his study. He tried to understand the effect of these sound waves on height and weight of cowpea seedlings. He treated cowpea plants with sound waves of six different types and frequencies. Treated plants were taller than control. All the different frequencies *viz.* 400 Hz frequency sound, cuckoo acoustic song and cricket acoustic song, followed by insect music mixed sound, Fn and F5 contributed to an improvement in growth. However, cuckoo and cricket acoustic song treatment had a positive influence on the weight of cowpea seedlings.

Effect of chewing vibration on plants

Herbivore-plant interaction is an interesting area of study. The influence of chewing sound on plant defense responses has ecological significance. Reports say that vibrations caused by insect feeding can elicit chemical defenses (Appel and Cocroft, 2014). *Arabidopsis thaliana* (L.) rosettes pre-treated with vibrations caused by caterpillar feeding had higher levels of glucosinolate and anthocyanin defense molecules when subsequently fed upon by *Pieris rapae* (L.) caterpillars than did untreated plants. Plants also discriminated between vibrations caused by chewing and those caused by wind or insect song. Plants thus respond to herbivore-generated vibrations in a selective and ecologically meaningful way. A vibration signalling pathway would complement known pathways that rely on volatile, electrical, or phloem-borne signals. They suggested that vibration may represent a new long distance signaling mechanism in plant–insect interactions that contributes to systemic induction of chemical defenses.

Mechanism of sound stimulation on plants

The exact mechanisms behind sound responses in plants are not well understood. However possible mechanisms proposed by different scientists are explained below.

Vibration speeds up protoplasmic streaming

One of main reasons underlying sound responses of plants is that sound vibrations speeds up protoplasmic streaming (Gagliano, 2013). Cytoplasmic streaming helps in the movement of proteins, nutrients and other organelles within and throughout the cell. The process is pH dependent, occurs along actin filament in the cytoskeleton of cell and is mediated by motor protein known as myosin which uses ATP as energy. It works in a manner that it tows the organelles and cytoplasm contents in the same direction. Sound in fact acts as an alternate stress on plants there by speeding up protoplasmic streaming which results in phase transition of protoplasm from colloid to gel. It increases the rate of metabolic activities which ultimately leads to better growth and development of plants.

Regulates plasmalemma H^+ ATPase activity

H^+ ATPase or proton pump regulates cellular membrane potential. It extrudes protons from cells of plants to generate electrochemical potential gradient. Generation of this gradient has a major role in providing energy for secondary active transport across membranes. It plays an important role in adaptation of plants to changing environment especially during stress condition. It has been reported that sound vibrations regulate proton pump activity (Yi *et al.*, 2003). It is also responsible for solute transport (root nutrient uptake) having a major role in cell growth.

Leaf canopy surface vibration increases rate of transpiration

Transpiration is the process of water movement through a plant and its evaporation from aerial parts. Transpiration occurs through stomatal apertures, and can be thought of as a necessary "cost" associated with opening of stomata to allow diffusion of carbon dioxide gas from air for photosynthesis. Transpiration also cools plants, changes osmotic pressure of cells, and enables mass flow of mineral nutrients and water from roots to shoots. In still air, water lost due to transpiration can accumulate in the form of vapor close to the leaf surface. This will reduce rate of transpiration, as water potential gradient from inside to outside of the leaf is lower. According to Hassanien *et al.,* 2014, sound vibrations will remove still air on the boundary layer of leaves there by increasing the rate of transpiration. Enhanced rate of transpiration will in turn speed up CO_2 diffusion in to the plant system which will result in higher photosynthetic rate and better absorption of nutrients from soil, ultimately leading to increased dry matter production.

Conclusion

The use of sound to improve plant health is a novel idea, which can have a significant impact on growth and development of plants. Whether the impact is stimulating or detrimental depends on the type of sound being played, exposed plant species and exposure stage of plant growth. This technology can be utilized for increasing yield and improving quality of produce. At present, application research on acoustic wave treatments on plants is still in experimental stage; data accumulation and constant exploration is needed. Standards should be fixed for the volume, type of acoustic frequency, species of plants and growth environment. However, this is a promising technology which requires further exploration as it ensures clean cultivation. Field experiments are needed to promote this technology for enhancing productivity of different crops.

References

Ankur, P., Sangeetha, A., and Seena, N. 2016. Effect of sound on the growth of plants: Plants pick up the vibrations. *Asian J. Plant Sci. Res.* 6(1): 6-9.

Appel, H. M. and Cocroft, R. B. 2014. Plants respond to leaf vibrations caused by insect herbivore chewing. *Oecologia* 175: 1257-1266.

Bochu, W., Jiping, S., Biao, L., Jie, L., and Chuanren, D. 2004. Sound wave stimulation triggers the content change of the endogenous hormone of the Chrysanthemum mature callus. *Colloids Surf. B. Biointerface* 37: 107-112.

Bochu, W., Xin, C., Zhen, W., Qizhong, F., Hao, Z., and Liang, R. 2003. Biological effect of sound field stimulation on paddy rice seeds. *Colloids Surf. B. Biointerface* 32: 29-34.

Bose, J. C., 1926. The Nervous Mechanism of Plants. Longmans, Green and Co. Ltd., 39 Paternoster Row, London, New York, Toronto, Bombay, Calcutta and Madras, Chapter VIII, pp 63.

Chivukula, V. and Ramaswamy, S. 2014. Effect of different types of music on *Rosa chinensis* plants. *Int. J. Environ. Sci. Dev.* 5(5): 431-434.

Chowdhary, E. K., Lin, H., and Bae, H. 2014. Update on the effect of sound wave on plants. *Res. Plant Dis.* 20(1): 1-7.

Chuanren, D., Bochu, W., Liu, W. Q., Jing, C., Jie, L., and Huan, Z. 2004. Effect of chemical and physical factors to improve the germination rate of Echinacea angustifolia seeds. *Colloids Surfaces B-Biointerfaces* 37: 101-105.

Creath, K. and Schwartz, G. E. 2004. Measuring effects of music, noise, and healing energy using a seed germination bioassay. *J. Alt. Complementary Med.* vol. 10(1): 113-122.

Ekici, N., Dane, F., Mamedova, L., Metin, I., and Huseyinov, M. 2007. The effects of different musical elements on root growth and mitosis in onion (*Allium cepa*) root apical meristem (musical and biological experimental study). *Asian J. Plant Sciences* 6(2): 369-373.

Fechner, G. T. 1848. *Nanna oder über das Seelenleben der Pflanzen*, Leipzig: Leopold Voß.

Gagliano, M. 2013. Green symphonies: a call for studies on acoustic communication in plants. *Behav. Ecol.* 24(4): 789-796. Available: http://beheco.oxfordjournals.org. [29 Jan. 2013].

Hassanien, H. E. R., Tian-zhen, H., Yu-feng, L., and Bao-ming, L. 2014. Advances in effects of sound waves on plants. *J. Integrative Agric.* 13(2): 335-348.

Haswell, E. S., Phillips, R., and Rees, D. C. 2011. Mechanosensitive channels: what can they do and how do they do it? *Structure* 19: 1356-1369.

Hou, T. Z., Luan, J. Y., Wang, J. Y., *et al.*, 1994. Experimental evidence of a plant meridian system: III. *Am. J. Chinese Med.* 22(3): 205-214.

Isono, K., Niwa, Y., Satoh, K., and Kobayashi, H. 1997. Evidence for transcriptional regulation of plastid photosynthesis genes in *Arabidopsis thaliana* roots. *Plant Physiol.* 114: 623-630.

Jun, H. and Shi-ren, J. 2011. Effect of six different acoustic frequencies on growth of Cowpea (*Vigna unguiculata*) during its seedling stage. *Agric. Sci. Technol.* 12(6): 847-851.

Lopez-Ribera, I. and Vicient, C. M. 2017. Drought tolerance induced by sound in *Arabidopsis* plants. *Plant Sig. Bev.* 12(10): e1368938- e1368938-7.

Meng, Q., Zhou, Q., Zheng, S., and Gao, Y. 2012. Responses on photosynthesis and variable chlorophyll fluorescence of *Fragaria ananassa* under sound wave. *Energy procedia* 16: 346-352.

Mishra, R. C., Ghosh, R., and Bae, H. 2016. Plant acoustics: in the search of a sound mechanism for sound signaling in plants. *J. Expt. Bot.* 67(15): 4483-4494.

Retallack, D. 1973. *The Sound of Music and Plants*. DeVorss in Santa Monica, California, 96p.

Seregin, I. V. and Ivanov, V. B. 2001. Physiological aspects of cadmium and lead toxic effects on higher plants. *Russ. J. Plant Physiol.* 48: 523-544.

Subramanian, S., Chandrasekharan, P., Madhava-Menon, P., Raman, V. S., and Ponnaiya, B. W. X. 1969. A study of the effect of music on the growth and yield of paddy. *Madras Agr. J.* 56: 510-516.

Weinberger, P. and Measures, M. 1979. Effects of the intensity of audible sound on the growth and development of Rideau winter wheat. *Can. J. Bot.* 57: 1036-1039.

Xiujuan, W., Bochu, W., Yi, J., Chuanren, D., and Sakanishi, A. 2003. Effect of sound wave on the synthesis of nucleic acid and protein in Chrysanthemum. *Colloids Surf. B. Biointerface* 29: 99-102.

Xiujuan, W., Bochu, W., Yi, J., Defang, L., Chuanren, D., Xiaocheng, Y., and Sakanishi, A. 2003. Effects of sound stimulation on protective enzyme activities and peroxidase isoenzymes of Chrysanthemam. *Colloids Surf. B. Biointerface* 27: 59-63.

Yi, J., Bochu, W., Xiujuan, W., Chuanren, D., and Xiaocheng, Y. 2003. Effect of sound stimulation on roots growth and plasmalemma H- ATPase activity of Chrysanthemum. *Colloids Surf. B. Biointerface* 27: 65-69.

Zakariya, F. H., Rivai, M., and Aini, N. 2017. Effect of Automatic Plant Acoustic Frequency Technology (PAFT) on Mustard Pakcoy (*Brassica rapa* var. *parachinensis*) Plant Using Temperature and Humidity Parameters. *Int. Seminar Intelligent Technol. Appln.* 359-364.

Zhao, H. C., Zhu, T., Wu, J., and Xi, B. S. 2002. Role of protein kinase in the effect of sound stimulation on the PM H+-ATPase activity of Chrysanthemum callus. *Colloids Surfaces B-Biointerfaces* 26: 335-340.

3

Spectral Manipulation of Plant Responses

Shafeeqa T, Nandini K* and *Girija T
Department of Plant Physiology, College of Horticulture
Kerala Agricultural University, Vellanikkara, Thrissur, Kerala

Introduction

The ultimate aim of agricultural research or any other related activity is to increase yield. Yield improvement could be obtained mainly through the enhancement of photosynthetic efficiency. Photosynthesis is a process in which plants convert light energy to chemical energy. Sun is the ultimate source of energy for photosynthesis. However only half of the total solar radiation (51%) reaches the earth's surface of these only 1.2% reaches the plant canopy. Solar spectrum contains an array of other radiations which can be harmful for plant and human life. With increasing global climatic change amount of harmful radiation reaching the earth's surface has increased and this is found to influence plant productivity. Under this scenario, protected cultivation is gaining popularity in many countries as it helps to reduce harmful effects of undesirable radiation, to allow cultivation of crops irrespective of seasons and also expand cultivation to non-traditional areas by artificially manipulating crop environment. In this context, spectral manipulation for crop production has emerged as an important area in crop production system

Light spectrum

Quantum theory envisages light as packets of energy known as photon or quanta. Solar radiation reaching the earth's surface contains the following radiations. Based on the energy levels of these radiations they can be arranged as gamma, X-ray, ultra violet, visible, infrared, micro wave and radio wave in the order of increasing wavelength (Table 1). Among these, radiations in the range of y 400- 700 nm alone can be absorbed by plants which is known as photosynthetically active radiation (PAR) or visible spectrum.

Table 1. Wavelength, frequency and energy level of the visible spectrum

	Wavelength	Frequency	Photon energy
Violet	380–450 nm	668–789 THz	2.75–3.26 eV
Blue	450–495 nm	606–668 THz	2.50–2.75 eV
Green	495–570 nm	526–606 THz	2.17–2.50 eV
Yellow	570–590 nm	508–526 THz	2.10–2.17 eV
Orange	590–620 nm	484–508 THz	2.00–2.10 eV
Red	620–750 nm	400–484 THz	1.65–2.00 eV

Absorption spectrum and action spectrum

Absorption spectra indicates how much each color of light is absorbed by various pigments of plants. It can be represented graphically to explain the quantum of light (electromagnetic radiation) of different wavelengths absorbed by a pigment after a single exposure to full light.

Action spectra describes the colors of PAR that are doing the actual work of driving photosynthesis. KJ McCree (1972) studied 22 different plant species and has reported that the photosynthetic active radiation (PAR) was between 400 and 700 nm indicating that maximum photosynthesis occurs within this range and among these wavelengths, red (600 to 700 nm) and blue (400 to 500 nm) regions are used by plants to create energy to drive the process of photosynthesis. Among these, the efficiency of red region is found to be higher this may be because in red region all the energy absorbed by the plant can be converted to chemical energy, but since blue region has higher energy level and lower wavelength complete transformation to chemical energy may not be taking place.

Table 2. Effect of different wavelengths of light on physiological processes of plants (Malik and Srivastava, 2000)

Wavelength (nm)	Physiological processes
280 and below	Lethal to plants and other micro organisms
280-400	Rosette growth: thick leaves
400-510	Active photosynthesis
510- 610	Least amount of photosynthesis and morphogenesis
610-700	Highest photosynthetic activity, pollen and seed germination, flowering, etc.
700-800	Pfr absorbs radiant energy

Properties of light

Light is the most familiar part of the electromagnetic spectrum which travels in straight line with a speed of 3,00,000 km/s in space. It can be absorbed, reflected and transmitted by objects on earth. The average speed of light will be less when it interacts with absorbing and re-emitting particles of the atmosphere. Due to its dual nature light can be treated either as a wave or particle. The greater the number of interactions along the light's path, the less the average speed. Plants respond to intensity, quality, duration and periodicity of light received by them.

i. Light intensity
ii. Light quality
iii. Light duration
iv. Light direction

Light intensity

Light intensity is a critical environmental factor that influences crop physiology and biochemistry. It is the amount of light reaching the plant canopy. Variation in light intensity can lead to considerable changes in leaf morphology and structure. According to Wu *et al.,* 2017, crop plants produce smaller and thinner leaves under low light conditions than corresponding leaves in full sunlight conditions. Dry matter of roots, stems, leaves, and whole plant as well as photosynthetic rate, transpiration and stomatal conductance, and stem diameter was found to be affected by the intensity of light (Yang *et al.,* 2014, 2017). While low light intensity may be insufficient for the synthesis of photoassimilates, high light intensity can cause serious damages to the plants, like destruction of chlorophyll, protein, DNA etc.

Light quality

The spectral quality of light has multivarient effects on plant growth and development. A number of studies have revealed the response of plants to different colour regimes of the spectrum. This has resulted in large scale use of LEDs in protected cultivation. Though all the spectral colours influence growth the amount of red and blue light reaching on the plant canopy is reported to have a strong influence on Calvin cycle enzymes like RuBP carboxylase phosphatase. Blue light is known to induce plant movements, regulate cicardian clock in plants and stomatal conductance, in addition to relocation of chloroplasts such as the bending of some structures towards light (Briggs and Christie, 2002; Spalding and Folta, 2005). Green light has also been reported to stimulate phototropic responses and leaf inclination perceived by the complementary light receptor, heliochrome (Steinitz *et al.,* 1985). Strong ultra violet radiation cause denaturation of protein as well as destruction of chlorophyll.

Light duration

Based on the photoperiod plants are classified in to long day plants, short day plants and day neutral plants. Long day plants require longer light period in a 24 h cycle for subsequent flowering, but short-day plants require a relatively short day light and continuous dark period of about 14-16 h for subsequent flowering, and day neutral plants flower in all photoperiods ranging from 5 h to 24 h continuous exposure.

Light direction

In nature due to changes in sun's angle, cloud cover, shading from overlapping leaves and neighboring plants, light intensities and spectral properties experienced by plants may vary. Such spatial and temporal gradients in incident light, can have major consequences for photosynthetic carbon assimilation in leaves (Pearcy, 1990; Chazdon and Pearcy, 1991; Pearcy and Way, 2012). This can also contribute to anatomical and morphological variations in the distribution of stomata as is seen in plants like *Sansevieria trifasciata* with vertical leaf have unifacial leaf anatomy with uniform distribution of stomata on both the leaf surfaces while in plant species like *Mangifera indica* with horizontal leaf arrangement stomata distributed will be more either on upper or lower side.

The direction of light can also influence movement of chloroplasts. It is observed that when the intensity is high, chloroplast move sideways and become parallel to the light source to reduce the harmful effects of high light intensity. Similarly, when light intensity is low, chloroplasts move perpendicular to the light source to absorb more light.

Light perception by plants

Light signals are absorbed by various proteinaceous pigment molecules called photoreceptors.

- There are at least three photoreceptors in plants to perceive information about their light environment, each specifically absorb different spectral ranges: red:far-red (R:FR)-sensing phytochromes, blue/UV-A photoreceptors, and UV-B photoreceptors (Kendrick and Kronenberg, 1994). All photoreceptors consist of proteins bound to light absorbing pigments i.e. chromophores.
- Spectral sensitivity of each photoreceptor depends on its chromophore's ability to absorb different wavelengths i.e. on the chromophore´s absorption spectrum.
- In response to light absorption, downstream signaling is mediated by the photoreceptor protein.
- Phytochromes can detect both red (Pr) and far-red (Pfr) light (R:FR: 665-670/ 730-735 nm). It consists of a kinase base with two identical proteins joined to form one functional unit. The kinase base is bound to a non-protein part known as a chromophore, which is the light absorbing part. The chromophore reverts back and forth between two isomeric forms, with one (Pr) absorbing red light and becoming Pfr, and the other (Pfr) absorbing far-red light and becoming Pr. Interconversion between isomers acts as a switching mechanism that controls various light-induced events in the life of plants. They also provide important information to a plant such as length of the diurnal dark period, potential and actual shading by other vegetation and depth of immersion in water.
- Phytochromes are encoded by five related genes called PHYA-E.

In 1990s, molecular biologists analyzing *Arabidopsis* mutants found three completely different types of pigments that detect blue light. These are cryptochromes (for the inhibition of hypocotyl elongation), phototropins (for phototropism) and a carotenoid-based photoreceptor called zeaxanthin (for stomatal opening). In addition to these, photobiological studies in response to green light revealed the existence of a far-red/ green reversible receptor termed heliochrome.

Cryptochromes

These receptors respond to blue and UV-A radiations (Lin, 2000). They affect the circadian clock and also mediates light regulation of gene expression. Cryptochromes are encoded by CRY1 and CRY2. Cry1 is known as a major blue light receptor regulating light induced flavonoid biosynthesis and also *CHS* (chalcone synthase) genes in Arabidopsis (Cashmore *et al.,* 1999; Sancar, 2000).

Phototropins

Phototropins respond to blue light and were defined as compounds that inhibit plant gravitropic and phototropic responses and polar transport of the hormone auxin. PHOT1 and PHOT2 (formerly known as NPH1 and NPL1 respectively) are the two genes characterized by phototropins. Their common structural theme is that benzoic acid is ortho-linked to a second aromatic ring system. 1-N-naphthylphthalamic acid (NPA) is widely used in phototropic studies, but 1-pyrenoylbenzoic acid is the most active synthetic compound. The UV-B photoreceptors are yet to be extensively characterized.

Photomorphogenesis

Light affects many developmental and physiological processes of plants from seed germination, hypocotyl hook unfolding, permeability of cell membranes, rhizome and bulb formation, flowering, deciding the direction of growth, seed maturation and induction of dormancy.

Exposure to light triggers several major developmental and physiological events in plants. These include: growth inhibition and differentiation of the embryonic stem (hypocotyl); maturation of embryonic leaves (cotyledons); and establishment and activation of stem cells in the shoot and root apical meristems. Recent studies have linked a number of photoreceptors, transcription factors, and phytohormones to each of these events. Light energy absorbed by plants is converted as signals which will enter the nucleus; by transduction it will then either express or suppress certain genes that will create a response in plants.

Molecular physiology of photomorphogenic processes

White light is a combination of different spectral wavelengths. The responses in plants to this visible spectrum are responsible for many of the photomorphogenic processes with certain specific spectral signatures corresponding to its blue, green and red regions. Among these spectral wavelengths, red and blue colours are found to have a major role in activating physiological processes in plants. All these plant responses are coordinated via different spectra-specific photoreceptors discussed in the previous section. Blue light was reported to activate the synthesis of anthocyanin by the activation of transcription factors regulating genes encoding anthocyanin biosynthetic enzymes (Mol *et al.,* 1996). In the orchid *Paphiopedilum,* the guard cells were found to be devoid of functional chloroplasts. However, an active blue light receptor has been identified in the underdeveloped orchid chloroplasts which activates xanthophyll cycle, thereby accounting for the blue/green reversibility and red-light enhancement properties of blue light-specific opening of stomata (Talbott *et al.*, 2002). IAA synthesis and polar transport of IAA are reported to be influenced by red light receptors/phytochromes (Liu *et al* 2011). Activation of nitrate reductase under red light has been reported in *Diplotaxis tenuifolia* (L.) DC. This was seen even under low nitrogen conditions (Signore *et al.*, 2020).

UV-B and UV-C radiations reaching the earth's surface have been studied extensively and it is seen that these high energy radiations also affect photomorphogenesis in plants. UV-B is generally found to have an adverse effect on morphological attributes such as plant height, leaf area and tiller number as observed from studies in rice (Shafeeqa and Nandini, 2018) and *Gerbera* (unpublished data). Higher chlorophyll degradation has been observed under high UV-B conditions, lowering the chlorophyll content. Photosynthetic rate and stomatal conductance were also observed to be lower under high UV-B levels. Products of secondary metabolic pathway such as flavonoid and anthocyanin contents were found to be higher. Studies conducted by Hu *et al.,* 2020 have shown that UV-B radiation activates a *MdWRKY72* transcription factor pathway promoting anthocyanin biosynthesis in apple. In *Gerbera jamesonii*, supplementation of red, blue, green and yellow light as well as UV-B exposure resulted in variations in flower stalk length in comparison to the respective control conditions (unpublished data; Figure 1). In this background, it is very pertinent that manipulation of supplemental radiation exposures can help in achieving growth advantages in plants.

Fig. 1: Spectral manipulation in *Gerbera jamesonii*:
(a) Experimental layout; **(b)** Variation in plant height under the influence of supplemental LED light with different spectral colours; **(c)** Variation in plant height under the influence of non- UV-B and UV-B exposure. (*Image Courtesy and details*: Anil A S)

Need of light manipulation

Seasonal variation in environmental condition can be very drastic in temperate countries. This also includes variation in light quality and intensity .Studies also indicate that "End-of-day" (EOD) or evening light quality has a strong influence on morphology of ornamental seedlings because evening time red and far red light occupies a major portion compared to other light so it will trigger photoreceptors like phytochrome or cryptochrome and lead to change in the inter nodal length. This can be adopted as a non-chemical way to improve plant height. These studies indicate that to sustain productivity and improve the market quality of produce, light manipulation can play a major role; hence protected cultivation practices are gaining a lot of importance.

To improve productivity in polyhouses, in addition to temperature and humidity control, manipulating the light environment is also possible. Various sources of light like high pressure sodium lamp (HPS), incandescent lights, fluorescent tube lights etc. are being used in horticulture for photoperiodic control as well as for production purpose. Currently new sources of lights such as light emitting diodes (LED) are popular.

Methods of light manipulation

Spectral manipulation can be done by two methods viz., light emitting diode and colour shade nets.

Light emitting diode

Light emitting diodes are solid state semiconductor devices which produce a narrow spectrum of light which may be of single or multiple wavelengths. LEDs have been used in photobiology from 1980s onwards. LEDs have strong control over plant growth with reduced energy and chemicals. It can be configured as overhead panels bars and as intra canopy configuration.

Flowering is influenced by day length or photoperiod. Artificial short night has been provided to promote flowering of long day plants. In petunia under normal condition 9h of day length, will take 61 days to flower but if we artificially provide 16h day using LED light source it will flower within 34 days.

Tomato transplants grown under different LED illumination like UV, yellow, blue orange and white light were observed for changes in various morphological characters like plant height, total fresh weight, hypocotyl diameter and number of flowers in the first inflorescence. Plant height was found to be higher under yellow light illumination and all other parameters like total fresh weight, hypocotyl diameter and number of flowers in the first inflorescence was found to be higher under ultra violet radiation at 380 nm.

Most of the growth indices like relative growth rate (RGR), leaf area ratio (LAR) and specific leaf area (SLA) were found to be higher under UV exposure (380 nm) but net assimilation rate (NAR) was higher under red light. NAR has direct correlation with leaf photosynthesis and leaf weight ratio (LWR) and shoot-root ratio (SRR) were found to be higher under green light (520 nm).

Another important aspect of using LEDs is that we can generate specific wavelengths for special purposes and also vary their proposition as per requirement. Earlier studies indicate that red and blue are important for plant growth and development. Blue light with higher energy improves the nutritional quality of fruits and vegetables. It was also observed that a proportion of 75% red and 25% blue light had higher effect on growth of tomato plants as compared to 100% red light and white light. However, the response varied with the species, wherein, white light could not be replaced by any other wavelength as in case of wheat.

Tweaking and tuning

Fine tuning the light environment around the plant is based on the requirement which will increase the germination and flowering. The main advantage of this is that it reduces light stress to plants.

Table 3. Effect of blue and red LEDs on plant growth (Massa *et al.*, 2008)

Crop	Types of LED	Effect on plant growth
Strawberry	Blue, Red	Blue-increase yield Red-increase mallic acid and oxalic acid Red + Blue-Firmness, shelf life, anthocyanin and glucose content
Sugar cane	Green	Increase growth, side offshoots, glucose content
Cotton	Blue, Red	Red + Blue-Increase shoots fresh and dry weights, nodal Distance, leaf thickness, chlorophyll contents
Wheat	Red	High morphogenesis, photosynthesis and seed yield
Tomato	Red, Far-Red	Increase seedling germination, yield, chlorophyll and systemic disease resistance against root knot nematode
Cucumber	Blue, Red	Improves growth, seed quality, Physiological and biochemical characteristics

As compared to other light sources, LEDs are small in size, have high durability, long operating lifetime, wavelength specificity, relatively cool emitting surfaces, linear photon output with electrical input current. Plants growing under LED light have higher growth compared to plants growing under high pressure sodium lamps. These advantages, coupled with new developments in wavelength availability, light output, and energy conversion efficiency has in fact initiated a revolution in horticultural lighting. Light emitting diode (LED) technology has emerged and developed rapidly in the past decades as an alternative light source.

Large scale use of LED has reduced the cost of LED lamps which has renewed interest in the use of LEDs as a tool in greenhouse research. Moreover, use of light emitting diode accelerates anthocyanin accumulation higher chlorophyll content in grape skin. Stem and pedicel length of grapes was also shorter in LED grown plants compared to plants grown under high pressure sodium lamp (HPS).

Colour shade nets

Coloured shade netting has become popular for manipulating plant growth and development. These nets are designed specifically for such uses and can be used both outdoors as well as in green houses. They provide physical protection from birds, snails, insects and also influence the micro climate such as humidity, temperature and light quality.

Colour nets and shading

Shading properties of the different coloured nets vary. Black colour net is having higher shading percentage (55% to 60%) followed by blue (51% to 57%), pearl (52% to 54%), red (41% to 51%) and green (40%- 50%).

Light absorption and transmission is found to vary with colour of the nets (Table 4). Spectral manipulation for controlling morphogenesis in plants is possible with the use of shade nets. The approximate transmission rate of PAR and spectral wavelength under different colours of shade net is given in Table 5. However, actual values will vary with shading percentage of the nets used.

Table 4. Light quality modification by color of shade nets (Yang *et al.*, 2012)

Net	Absorption	Transmittance
Blue	UV+Y+R+FR	B+G
Red	UV+B+G	R+FR
Yellow	UV+B	G+Y+R+FR
White	UV	B+G+Y+R+FR
Pearl	UV	B+G+Y+R+FR
Grey	all+IR	-
Black	all	-

UV- Ultraviolet; B- Blue; G- Green; Y- Yellow; R-Red, FR- Far Red

Table 5. Transmittance rate under different coloured shade nets

Colour	PAR shading (%)	Transmitted light at PAR (%)
Black	80	20
Blue	79	21
Gray	76	24
Red	68	32

Factors to be considered while using shade nets

1. **Radiation:** Reduction in radiation resulting from netting will affect temperature and relative humidities.
2. **Radiation scattering:** Diffuse light has been shown to increase radiation use efficiency and also affects plant flowering
3. **Photoselectivity:** Colored shade nets can be used to change red to far-red light ratio that are detected by phytochromes.

4. **Air movement:** netting also reduce wind speed and wind run which can affect temperature relative humidities and gas concentration resulting from reduction in air mixing.
5. **Temperature:** Shade nets are often deployed over crops to reduce heat stress.'
6. **Relative humidity:** Relative humidities are often higher under netting than outside as a result of water vapor being transpired by the crop and reduced mixing with drier air outside the netted area.

Benefits of shade netting

Shade nets distribute light uniformly over canopy of the crop, because they serve as an obstacle to direct light and cause diffusion which will reduce adverse effects of high light intensity like sun scorching and sunburn.

A large percentage of solar radiation is scattered under green colour shade net followed by blue, red, grey and black. Green colour shade nets are more preferred under high light intensity which will reduce the intensity by scattering more light. Red nets have an optimum range of scattering and is more permissible for red and far red light. Since phytochrome activity will be more under red nets, they are widely used for better results under protected condition.

Shade net influence on plant characters

A number of studies have confirmed the effect of different coloured nets on plant phenology, growth and quality of fruits and vegetables. Red netting in comparison with black and blue netting was reported to induce earlier flowering of orchid *Phalaenopsis* cultivars and hybrids. Fresh weight of harvested leaves and total number of harvestable leaves of variegated cast iron plant was found to be higher under black netting than under blue, grey and red. In *Pittosporum variegatum,* stimulation of vegetative growth by red net, blue net induced dwarfing and black net enhanced branching were observed. Tomato plants under different coloured shade nets possessed higher chlorophyll a, b and carotenoid contents under black net, whereas leaf area index and vegetative growth of plants and lycopene content improved under red net as compared to open condition (Shahak, 2000).

Table 6. Crop response to colour of shade nets (Shahak, 2000)

Crops	Colour nets	Effect on plants
Foliage and vegetables	Yellow and red	Vegetative growth
	Blue	Dwarfing
	Grey	Branching and bushiness
Cut flowers	Red and yellow	Longer and thicker flowering stems
	Blue	Dwarfing
	Red	Early flowering
	Pearl	Enhance the branching
Fruit crops	Red and pearl	Quality

Conclusion

Light is a major resource for sustaining plant life on earth. Manipulating spectral quality and intensity for improving crop production and quality improvement is gaining a lot of attention. Availability of affordable materials in the market such as LED lamps and shade nets has helped to popularize the technology. Moreover, these technologies help to improve productivity and quality without any harzardous effect on the crop or the environment and promote clean cultivation.

References

Brazaityte, A., Duchovskis, P., Urbonaviciute, A., Samuoliene, G., Jankauskiene, J., Sakalauskaite, J., Sabajeviene, G., SIrtautas, R., and Novickovas, A. 2010. The effect of light emitting diodes lighting on the growth of tomato transplants. *Zemdirbyste Agric.* 97(7): 89-98.

Briggs, W. R. and Christie, J. M. 2002. Phototropins 1 and 2: versatile plant blue-light receptors. *Trends in Plant Science* 7:204– 210.

Cashmore, A. R., Jarillo, J. A., Wu, Y. J., and Liu, D. 1999. Cryptochromes: Blue light receptors for plants and animals. *Science* 284: 760–765.

Chazdon, R. L. and Pearcy, R. W. 1991. The importance of sunflecks for forest understory plants. *Bioscience.* 41: 760–766.

Darko, E., Heydarizadeh, P., Schoefs, B., and Sabzalian, M.R. 2014. Photosynthesis under artificial light: the shift in primary and secondary metabolism. *Philosophical Trans. R. Soc.* 369: 1-7.

Goinso, C. D., Yorio, N. C., and Sanwo, M. M. 1997. Photomorphogenesis, photosynthesis, and seed yield of wheat plants grown under red light- emitting diodes (LEDs) with and without supplemental blue lighting. *J. Exp. Bot.* 48(312): 1407- 1413.

Hu, J., Fang, H., Wang, J., et al., 2020. Ultraviolet B-induced MdWRKY72 expression promotes anthocyanin synthesis in apple. *Plant Science: an International Journal of Experimental Plant Biology.* 292: 110377.

Illic, Z. S., Milenkovic, L., Sunic, L., and Falik, E. 2012. Effect of modification of light intensity by colour shade nets on yield and quality of tomato fruits. *Hortic. Sci.* 139: 90-95.

Illic, Z. S., Milenkovic, L., Sunic, L., and Falik, E. 2015. Effect of coloured shade-nets on plant leaf parameters and tomato fruit quality. *J. Sci. Food Agric.* 95: 2660-2667.

Kendrick, R. E. and Kronenberg, G. H. eds., 2012. *Photomorphogenesis in plants.* Springer Science & Business Media.

Liu, X., Cohen, J. D., and Gardner, G. 2011. Low-fluence red light increases the transport and biosynthesis of auxin. *Plant Physiology 157*(2): 891–904.

Massa, G. D., Kim, H. Y., Wheeler, M., and Mitchell, C. A. 2008. Plant productivity in response to LED lighting. *Hortic. Sci.* 43(7): 1951-1956.

McCree, K. J. 1972. The action spectrum, absorptance and quantum yield of photosynthesis in crop plants. *Agric. Meteorol.* 9(3–4):191–216.

Mol J, Jenkins GI, Schäfer E, Weiss D. 1996. Signal perception, transduction and gene expression involved in anthocyanin biosynthesis. *Critical Reviews in Plant Science* 15: 525– 557.

Oren-Shamir, M., Gussakovsky, E. E., Shpiegel, E., and Shahak,Y. 2001. Coloured shade net can improve the yield and quality of green decorative branches of *Pittosporum variegatum.J. Hortic. Sci. Biotechnol.* 76: 353-361.

Pearcy, R. W. 1990. Sunflecks and photosynthesis in plant canopies. *Annu. Rev. Plant Physiol. Plant. Mol. Biol.* 41: 421–453.

Pearcy, R. W. and Way, D. A. 2012. Two decades of sunfleck research: looking back to move forward. *Tree Physiol.* 32: 1059–1061.

Sancar, A. 2000. Cryptochrome: the second photoactive pigment in the eye and its role in circadian photoreception. *Annu. Rev. Biochem.* 69: 31–67.

Shafeeqa, T. and Nandini, K. 2018. Effect of ecofriendly chemicals on morphological and phenological characters of rice variety Uma under UV stress. *Agricultural Science Digest* (38): 172-177.

Shahak, Y. 2008. Photo selective netting for improved performance of horticultural crops. *Acta Hortic.* 770:161-168.

Signore, A., Bell, L., Santamaria, P., Wagstaff, C., and Van Labeke, M-C. 2020. Red Light Is Effective in Reducing Nitrate Concentration in Rocket by Increasing Nitrate Reductase Activity, and Contributes to Increased Total Glucosinolates Content. *Front. Plant Sci.* 11: 604.

Spalding, E. P. and Folta, K. M. 2005. Illuminating topics in plant photobiology. *Plant, Cell and Environment.* 28: 39-53.

Stamps, R. H. 2009. Use of coloured shade netting in horticulture. *Hortic. Sci.* 44(2): 239-241.

Steinitz, B., Ren, Z. L., and Poff, K. L. 1985. Blue and green light-induced phototropism in *Arabidopsis thaliana* and *Lactuca sativa* L. seedlings. *Plant Physiology.* 77: 248-251.

Stutte, G. W. 2008. Light emitting diodes for manipulating the phytochrome apparatus. *Hortic. Sci.* 44(2): 79-82.

Talbott, L. D., Zhu, J., Han, S. W., and Zeiger, E. 2002. Phytochrome and blue light-mediated stomatal opening in the orchid, paphiopedilum. *Plant Cell Physiol.* 43(6): 639-646.

Wu, Y., Gong, W., and Yang, W. 2017. Shade Inhibits Leaf Size by Controlling Cell Proliferation and Enlargement in Soybean. *Sci. Rep.* 7:9259.

Yang F., Huang S., Gao R., Liu W., Yong T., Wang X., et al., 2014. Growth of soybean seedlings in relay strip intercropping systems in relation to light quantity and red:far-red ratio. *Field Crop Res.* 155:245–253.

Yang F., Liao D., Wu X., Gao R., Fan Y., Ali Raza M., et al., 2017. Effect of aboveground and belowground interactions on the intercrop yields in maize-soybean relay intercropping systems. *Field Crop Res.* 203:16–23.

Yang, X., Wang, X., Wang, L., and Wel, M. 2012. Control of light environment: a key technique for high yielding and high quality vegetable production in protected farmland. *Agric. Sci.* 3: 923-928.

4

Geomagnetic Responses in Plants

Sreepriya S, Girija T* and *Parvathi M S

Department of Plant Physiology, College of Horticulture
Kerala Agricultural University, Vellanikkara, Thrissur, Kerala

Introduction

Earth's magnetic field (MF) is a natural component of environment and is an inescapable factor for all living organisms including plants. Earth has an inner core made of molten iron which rotates at a fairly high speed creating a magnetic field just like a dynamo. This Geo Magnetic Field (GMF) extends from the earth's interior into outer space. It interacts with the solar wind, which is actually a stream of charged particles emanating from the sun. As earth is engulfed in a GMF, all living organisms, both plants, animals and all biological processes are under its influence. However, local differences in strength and direction of the geomagnetic field have been detected. The vertical component of GMF is maximum at the magnetic poles, which is around 67 μT, and the horizontal component is zero there. Conversely, at the magnetic equator the vertical component is zero and the horizontal component is maximum, about 33 μT (Kobayashi *et al.*, 2004).

Turbulent flows of the fluid metallic inner core of earth mainly contribute to GMF and the effect of external MFs located in the ionosphere and the magnetosphere is very meagre (Qamili *et al.*, 2013). As GMF extends beyond the earth's surface it is this force along with magnetosphere that protects the earth and its biosphere, from the solar wind by deflecting most of its charged particles (Occhipinti *et al.*, 2014).

Geomagnetic field reversal and angiosperm evolution

Since GMF has been present since the beginning of plant life along with gravity, light, temperature and water availability it has also played a role in plant evolution. Among these factors except gravity, all others, including GMF, changed consistently during plant evolution contributing to abiotic stress and selection pressure eventually resulting in plant diversification and speciation. Along the different eons, GMF has exhibited several changes of magnetic polarity due to geomagnetic reversal along with persistent periods of same polarity. They occurred some hundred times since earth's formation and the mean time between a reversal and the next one has been estimated around 300,000 years. According to De Santis *et al.*, (2004) an imminent geomagnetic reversal would not be so unexpected. One school of thought on evolution claims that during magnetic reversal, biological material on earth may have been exposed to more intense cosmic radiation or UV light as a result of which mutations may have occurred which may be the cause for high level speciation. As plants are sessile, terrestrial plants might have been the most severely affected groups (Tsakas and David, 1986).

The oldest Angiosperm fossils date from early Cretaceous period, 130-136 million years (Myr) ago, followed by a rise to ecological dominance in many habitats before the end of Mesozoic era. It has been shown that periods of normal polarity transitions overlapped with diversion of most of the familiar angiosperm lineages (Fig.1). This correlation appears to be particularly relevant to angiosperms compared to other plants (Occhipinti *et al.,* 2014). However, it appears that many evolutionary changes in angiosperm families were either preceding or succeeding changes in GMF polarity patterns. Most of the plant families evolved in normal (N) polarity which were preceded and succeeded by normal polarity conditions (N-N-N type) whereas few others evolved in normal polarity either preceded and/or succeeded by normal or reversed (R) polarity patterns (R-N-N or R-N-R types), among which interestingly the recently evolved Orchidaceae belonged to N-R-N type, which is also depicted in Fig. 1.

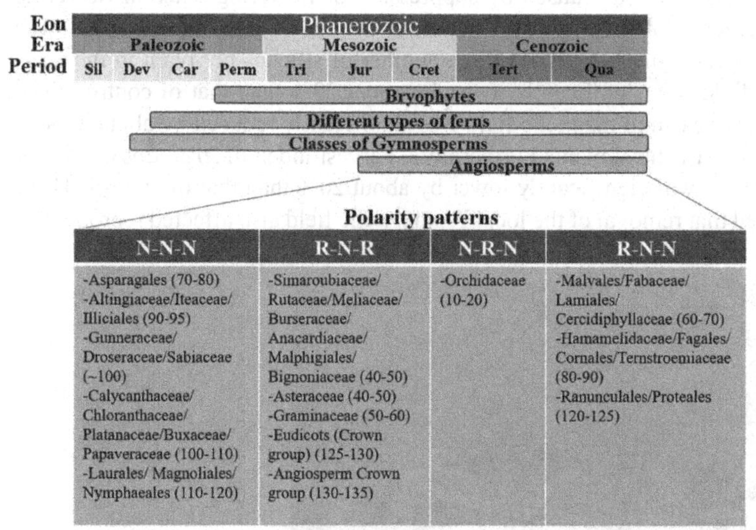

Fig. 1: Geomagnetic field reversals and Angiosperm evolution.

N-N-N: Evolution during normal (N) polarity preceded and succeeded by N; R-N-N: Evolution during N preceded by reverse (R) polarity and succeeded by N; R-N-R: Evolution during normal polarity, N preceded and succeeded by reversal in polarity, R; N-R-N: Evolution during R preceded and succeeded by N; Other abbreviations: Myr, million years ago; Jur., Jurassic; Perm., Permian; Quat., Quaternary; Tert., Tertiary; Tri., Triassic. Values in parenthesis indicate their evolution in Myr ago. (Adapted from Occhipinti *et al.,* 2014).

The evolutionary history of several plant groups varied across the different periods corresponding to Paleozoic (>250-430 Myr ago), Mesozoic (>130-250 Myr ago) and Cenozoic (0->130 Myr ago) eras, with Angiosperms being the latest evolved members. It is interesting to note that among Angiosperms, most of the evolutionary diversions occurred during periods of normal magnetic polarity either preceding or succeeding reversal in polarity patterns.

Effect of GMF on reproductive growth of plants

Effect of removing the geomagnetic field on growth of *Arabidopsis* was investigated by Xu *et al.,* (2013). He designed an equipment to grow plants under laboratory conditions in a near-null magnetic field for the whole growth period. The equipment possesses three mutually perpendicular pairs of Helmholtz coils and three sources of high-precision direct current power for the generation of a near-null magnetic field (Fig. 2). Studies showed that the biomass of *Arabidopsis* seedlings grown under near-null magnetic field did not show significant difference compared to control at vegetative stage. However, when fresh weight of *Arabidopsis* plants was measured 35 days after germination, plant biomass in the near-null magnetic field was significantly lower by about 36% than that of plants grown in the local geomagnetic field. At later growth stages biomass of *Arabidopsis* plants did not show any significant difference between magnetically treated plants and control plants.

During reproductive stage, a significant inhibition in growth by removal of the local geomagnetic field was caused by suppression of flowering wherein flowering time was delayed by about 5 days in near-null magnetic field compared to those grown in local geomagnetic field. Average number of siliques per plant in the near-null magnetic field was significantly lower by about 22% than that of control. Seed yield of plants in near-null magnetic field was significantly reduced by about 81% as compared to that of control plants.Furthermore, harvest index of *Arabidopsis* in near null magnetic field was significantly lower by about 20% than that of control. These results indicated that removal of the local geomagnetic field also affected reproductive growth of *Arabidopsis*.

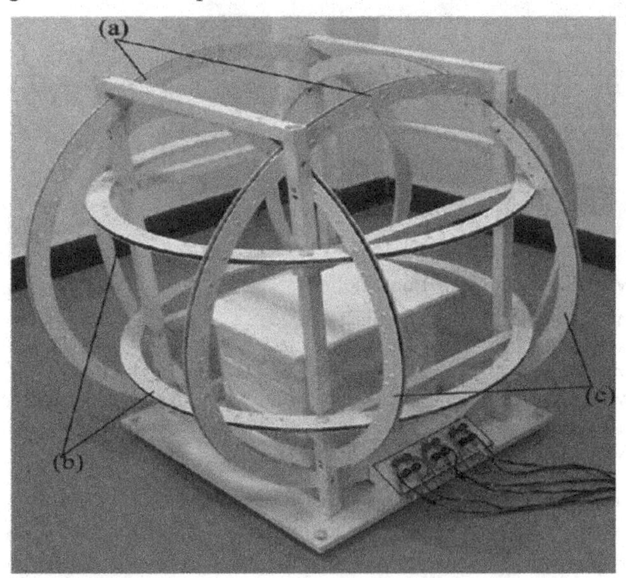

Fig. 2: Near-null magnetic field generating equipment: a,b,c are three intersecting coils carrying current in three different planes creating the required magnetic field in the centre (Xu *et al.*, 2013)

Ultrastructure changes of cells by exposure to low magnetic field

Exposing pea roots to Low Magnetic Field (LMF) of 0.5-2 nT showed changes in the ultrastructure of meristem cells as compared to control (Belyavskaya, 2001). Significant alterations were visible in the organelles of meristem cells as described below.

Nucleus: Normally, the chromatin is uniformly distributed in the nucleus, but after exposure to LMF it was found to be associated with the nuclear envelope. A reduction in rRNA synthesis was noticed which was probably due to reduction in the volume of granular nucleolus component and appearance of nucleolus vacuoles in LMF-exposed cells.

Plastids: Phytoferretin is an iron protein complex with a hollow sphere with 24 subunits which can accommodate 4500 iron atoms per molecule inside its internal cavity (Harrison and Arosio, 1996). It is normally seen spread in a thin array on the plastoglobuli of plastids and freely distributed in stroma in all of the sectional areas of plastids. It helps to store iron in a nontoxic form in plants. It was reported that 14% of plastids of meristem cells of control plants had electron-dense inclusions of phytoferritin with a diameter near 7 nm, while in LMF-exposed cells, only 1.5% of plastids had phytoferritin. Moreover, the granule number did not exceed 10-20 per plastid section. Such a low level of phytoferritin in plastids of LMF treated plants suggested that LMF treatment has resulted in either repression of phytoferritin synthesis or rise in its utilization for synthesis of iron-containing proteins.

Mitochondria: LMF exposure contributed to significant alterations in structure of mitochondria. Number of mitochondria per cellular section (population density) increased by 12% and diameter of mitochondria was about 1.5-2.0 fold more than in control cells. LMF exposure changed the structural configuration of the organelle; those with elongated shape under normal condition (control) became roundish. Changes were also observed in the number and location of cristae. They were seen as narrow short tubes located on organelle periphery. Number of cristae decreased significantly and they were absent in some organelles. Mitochondrial matrix was prominent with electron-dense inclusions whose structure was not specific. Mitochondrial ribosomes were rarely found in organelle sections and filaments of mitochondrial nucleoids were also absent. Moreover, since mitochondria are known as Ca^{2+} buffer that can pump out excess Ca^{2+} from cytoplasm, cytochemical experiments were carried out to reveal Ca^{2+} localization by means of pyro-antimonite method. Deposits of Ca^{2+} pyroantimonate granules observed in periplasmic space and cell walls as well as in Ca^{2+} sequestering stores (nuclei, plastids and endoplasmic reticulum) were rarely encountered in meristem cells of stationary control.

Cell wall: Ca^{2+} deposits in cell walls was found to decrease, which was demonstrated by the disruption in Ca^{2+} balance under LMF exposure, as revealed by cytochemical studies.

Effect of magnetic field on pollen germination

Betti (2011) studied the effects of a weak static magnetic field (MF) at 10 μT oriented downwards, combined with a 16-Hz sinusoidal MF (10 μT), on *in vitro* pollen germination of kiwifruit (*Actinidia deliciosa*). Standard pollen culture was done by suspending pollen in a standard liquid growth medium containing 0.29 M of sucrose, 0.4 mM of boric acid, and 1 mM of calcium nitrate. Two stress treatments were provided by suspending in 0.5 mM of calcium nitrate and 0 mM of calcium nitrate. In the fourth treatment instead of calcium, a weak static magnetic field (MF) at 10 μT oriented downwards, combined with a 16-Hz sinusoidal MF (10 μT) was given. The results indicated that magnetic field treatment partially removed the inhibitory effect caused by the lack of Ca^{2+} in culture medium, inducing a release of internal Ca^{2+} stored in the secretary vesicles of pollen plasma membrane. Although there is no direct information on the target components of weak electromagnetic exposure in "pollen system", calcium ions could constitute one of the possible targets, Ca^{2+} transport and binding being the parameters most consistently influenced by magnetic fields (Baureus, 2003).

Effect of GMF on root gravitropism

Kordyum *et al.,* (2005) studied root gravitropism in cress (*Lepidium sativum* L.) seedlings in the presence of weak, alternating magnetic field that consisted of a sinusoidal frequency of 32-Hz inside a metal shield. Gravistimulation of cress roots in the WCMF (Weak combined magnetic field) showed upward or negatively gravitropic roots whereas the control (GMF) showed normal positively geotropic roots. Since this frequency corresponded to the cyclotron frequency of Ca^{2+} ions, the observation suggested the participation of calcium ions in root gravitropism.

Light-microscopy showed difference in the pattern of amyloplast distribution in statocytes of gravistimulated roots in controls and in 32-Hz WCMF treatment. After 60 min of gravistimulation, all amyloplasts sedimented in the distal part of control columella cells. However, in the WCMF, they were localized close to one of the longitudinal walls; sometimes one or several amyloplasts touched the walls. The study indicated that a weak magnetic field along with oscillating magnetic field can change the rate and the direction of Ca^{2+} ion flux in roots, thereby affecting auxin distribution necessary for normal positive geotropism in roots.

Cryptochrome as magnetoreceptors

In plants, the blue light receptors, cryptochromes, have been suggested to act as a magnetoreceptor since they are involved in sensing the geomagnetic field. Cryptochromes in general influence several growth and developmental processes in plants by activating photochemical reactions. They play a role in de-etiolation responses such as inhibition of hypocotyl growth (Ahmad and Cashmore, 1993), production of anthocyanin (Ahmad *et al.,* 1995), expansion of leaf and cotyledons (Lin, 2002) and also transition to flowering. All these developmental processes in plants are controlled by blue-light regulated genes.

In *Arabidopsis*, cryptochromes are encoded by two similar genes, *cry1* and *cry2*. CRY2 protein levels in seedlings decreases rapidly upon illumination by blue light, as a result of protein degradation of light-activated form of the receptor. Structure of cryptochrome consists of N-terminal phytolyase homology domain which binds to flavin adenine dinucleotide cofactor and a C-terminal domain of variable length involved in signaling.

This was later confirmed by Xu *et al.*, (2012) by using his near-null magnetic field (NNMF) equipment. He used *Arabidopsis* as the test plant and found that in a NNMF, inhibition of *Arabidopsis* hypocotyl growth by light was weakened and flowering time was delayed. It also changed the expression levels of three cryptochrome-signaling-related genes *PHYB*, *CO* and *FT*. The transcript level of *PHYB* was elevated by 40%, and that of *CO* and *FT* was reduced by 40 and 50%, respectively.

Mechanism of magnetoreception

Ability of blue light receptors (cytochromes) to act as magnetoreceptors has been explained by 'radical pair model' based on the proposition that radical pairs are involved in magnetoreception. The blue light receptor, cryptochrome can form radical pairs after exposure to blue light which modifies the functions of cryptochromes (Occhipinti *et al.*, 2014). Flavin cofactor FAD of cryptochrome absorbs a photon by absorbing blue light, gets activated and becomes promoted to an excited FAD* state which further receives an electron from a nearby tryptophan, leading to the formation of $[FAD^{\bullet-}+TrpH^{\bullet+}]$ radical pair, which exists in singlet$^{(1)}$and triplet$^{(3)}$ states by coherent geomagnetic field-dependent inter conversions (Fig. 3). $FADH^{\bullet}$ can slowly revert back to the initial inactive FAD state and also through the inactive $FADH^{-}$ state of the flavin cofactor under aerobic condition (Fig. 3). The strength and direction of magnetic field affects the interconversion of singlet and triplet states which yields reactive oxygen species and act as signals in magneto sensing.

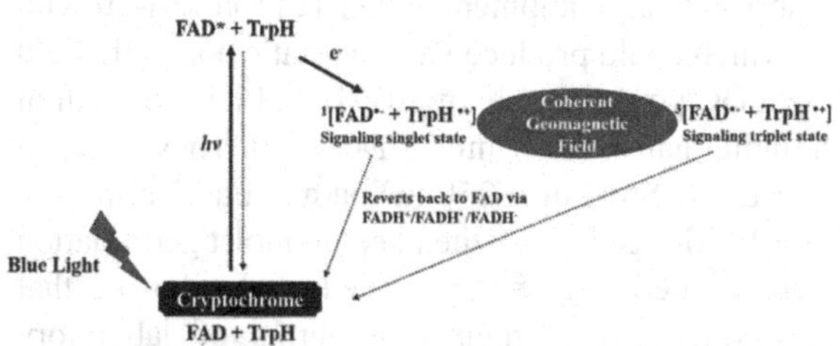

Fig. 3: Radical pair mechanism of magnetoreception. FAD: Flavin Adenine Dinucleotide, Trp: Tryptophan (Adapted from Occhipinti *et al.*, 2014)

Applications of plant geomagnetic responses

Magneto culture

Magneto culture is currently used to boost agricultural productivity. This technique taps earth's magnetic field and atmospheric electrical natural forces to boost plant growth, impart pest control and improve harvest quality and shelf life of crops. Earth's magnetic, paramagnetic, telluric and cosmic forces are used to balance the energetics of soil, water and plant growth which improves productivity. Generally, magnetic antenna, magnetite or permanent magnets are installed in the field along the north-south direction in a trench. The beneficial effects of magneto culture has been successfully demonstrated in carrot, sunflower and cabbage (www.electroculturevandoorne.com).

Magnetized water for irrigation

Normal water molecules are attracted by Van der waals force, which include attraction and repulsions between atoms, hence molecule clusters comprising of many water molecules are loosely attracted. This loose and chaotic form of attraction predisposes them to accumulate toxins and pollutants inside the water molecule cluster. These large water molecule clusters with toxins are blocked when they pass through cell membrane. However, when these chaotic clusters are smaller in size then, some of them carrying toxins, can enter the cell with consequent harmful effects (Florez, 2007). Therefore, to hydrate a plant, a great deal of normal water is required. Magnetic treatment of water restructures the water molecules into very small clusters, each made up of six symmetrically organized molecules. This tiny and uniform cluster has a hexagonal shape, so it can easily enter through passage ways in plant and animal cell membranes. In addition, toxic agents cannot enter its structure. It also exerts several effects on the soil-water-plant system (Ali *et al.,* 2014).

The impact of magnetized water application for improving common bean (*Phaseolus vulgaris* L.) production was studied by Moussa (2011). Control plants were irrigated with non-magnetized water and treated plants were irrigated with magnetized water of 30 mT. Growth characteristics, biochemical contents, antioxidant enzyme activity and physiological activity of crop was improved by the use of magnetized water, whereas no significant change was observed in hydrogen peroxide and lipid peroxidation product, malondialdehyde.

Pre-sowing magnetic treatment of seeds

Vashisht and Nagarajan, (2008) fabricated an equipment called Testron EM-20 with a gap of 5 cm between pole pieces which could produce variable static magnetic field (SMF) of 50-500 mT strength with a DC power supply of 80V/10A. Field strength in the pole gap was monitored by a digital gauss meter model DGM-30. They exposed seeds of chickpea (*Cicer arietinum* L.) to SMF of 0-250 mT in a cylindrical plastic sample holder for durations ranging 1-3 h. Seeds were then kept in moist germination paper and seedling characters were studied after 8 days. Their study showed that magnetic field application enhanced seed performance in terms of laboratory germination, speed of germination, seedling length, seedling dry weight and vigour index significantly compared to unexposed control. However, the response varied with field strength and duration of exposure without any particular trend. Among the various combinations of field strength and duration, 50 mT for 2 h, 100 mT for 1 h and 150 mT for 2 h exposures gave best results. In soil, seeds exposed to these three treatments produced significantly increased seedling dry weights in one-month old plants. Root characteristics of the plants showed dramatic increase in terms of root length, root surface area and root volume. Improvement in functional root parameters suggest that magnetically treated chickpea seeds may perform better under rainfed (non-irrigated) conditions where there is a restrictive soil moisture regime. Favourable effects of magnetic field on germination and emergence of seeds were also reported for cereals (Pietruszewski, 1999) and legumes (Podleoeny, 2003).

Weed suppression by magnetic field treatment

Differential response of plants to magnetic field was demonstrated by Balouchi and Sanavy in 2008. They used three *Medicago* species and dodder and treated them with different intensities of electromagnetic field and exposure times and studied the differences in seed germination characteristics. There was a significant decrease in germination percentage, seedling dry mass and seedling vigour index in dodder seeds, whereas in *Medicago* plants, germination percentages were improved by all treatments except 88 μT for 30 min in *M. scutellata.*

Exposure for 10 min to magnetic treatment improved shoot length of *M. radiata* and *M. polymorpha;* but this was not true for *M. scutellata.* However, in case of root length, all the treatments had higher root length than control. In *M. radiata,* 10 min exposure improved the root length but this was not true for *M. polymorpha.* These observations clearly indicated variations in varietal response to magnetic treatment. These studies also show that magnetoculture can be used as an organic tool for weed control.

Since magnetic field treatments decreased germination percentage and seedling vigour of *Cuscuta* and improved germination percentage and seed vigour of *Medicago sp.*, magnetic field treatments could be used as a pre-sowing treatment to control dodder weed population in the field and separate the seed lots of *Medicago*, which are contaminated with *Cuscuta* seeds.

Conclusion

Plant evolution is synchronized with geomagnetic field. GMF has influence on biological processes on earth; hence GMF is an inescapable environmental factor for plants on earth that affects plant growth and development. Elimination of GMF negatively affects normal plant development. Recent understanding of MF reversal with plant evolution opens new horizons not only in plant science but also to the whole biosphere, from the simplest organisms to human beings. Understanding the role of cryptochromes as magnetoreceptors in plants will provide the fundamental background necessary to understand evolution of life forms in our planet and will help us to develop and design scientific support systems in agriculture to utilize this knowledge in a beneficial way for improvement of crop productivity.

References

Ahmad, M. and Cashmore, A. R. 1993. Hy4 gene of *A. thaliana* encodes a protein with characteristics of a bluelight photoreceptor. *Nat.* 366:162-166.

Ahmad, M., Lin, C. T., and Cashmore, A. R. 1995. Mutations throughout an *Arabidopsis* blue-light photo receptor impair blue-light-responsive anthocyanin accumulation and inhibition of hypocotyl elongation. *Plant J.* 8:653-658.

Ali, Y., Rashidi, S., and Kavakebian, F. 2014. Applications of magnetic water technology in farming and agriculture development: a review of recent advances. *Curr. World Environ.* 9(3):695-703.

Balouchi, H. R. and Sanavy, M. 2009. Electromagnetic field impact on annual medics and dodder seed germination. *Int. Agrophys*. 23:111-115.

Baureus, K., Sommarin, M., Persson, B. R., Salford, L. G., and Eberhardt, J. L. 2003. Interaction between low frequency magnetic fields and cell membranes. *Bioelectromagnetics* 24:395-402.

Belyavskaya, N. A. 2001. Biological effects due to weak magnetic field on plants. *Adv. Space Res.* 34:1566-1574.

Betti. L., Trebbi, G., Fregola, F., Zurla, M., AND Brissi, M. 2011. Weak static and extremely low frequency magnetic fields affect *in vitro* pollen germination. *The Sci. World J.* 11:875–890.

De-Santis, A., Tozzi, R., and Gaya-Pique, L. R. 2004.Information content and K-entropy of the present geomagnetic field. *Earth Planet. Sci. Lett.* 218:269–275.

Florez, M., Carbonell, M. V., and Martínez, E. 2007. Exposure of maize seeds to stationary magnetic fields: Effects on germination and early growth. *Environ. Exp. Bot.* 59(1):68- 75.

Harrison, P. M. and Arosio,P. 1996. The Ferritins: molecular properties, iron storage function and cellular regulation. *Biochem. Biophys. Acta* 1275:161-203.

Kobayashi, M., Soda, N., Miyo, T., and Ueda, Y. 2004. Effects of combined DC and AC magnetic fields on germination of horn wort seeds. *Bioelectromagnetics* 25:552-559.

Kordyum, E. L., Bogatina, N. I., Kalinina, Y. M., and Sheykina, N.V. 2005. A weak combined magnetic field changes root gravitropism. *Adv. Space Res.* 36:1229-1236.

Lin, C. T. 2002. Blue light receptors and signal transduction. *Plant Cell* 14:207-225.

Moussa, H. R. 2011. The impact of magnetic water application for improving common bean (*Phaseolus vulgaris* L.) production. *N. Y. Sci. J.* 4(6):15-20.

Occhipinti, A., Santis, A., and Maffei, M. E. 2014. Magnetoreception: an unavoidable stepfor plant evolution? *Trends Plant Sci.* 19:1-4.

Pietruszewski, S. 1999. Magnetic treatment of spring wheat seeds. Rozprawy Naukowe, University of Agriculture Press, Lublin, Poland, 120p.

Podleoeny, J., Lenartowicz, W., and Sowinski, M. 2003. The effect of pre-sowing treatment of seeds magnetic biostimulation on morphological feature formation and white lupine yielding. *Zesz. Probl. Post. Nauk Roln.* 495:399-406.

Qamili, E., Santis, A., Isac, A., Mandea, M., Duka, B., and Simonyan, A. 2013. Geomagnetic jerks as chaotic fluctuations of the Earth's magnetic field. *Geochem. Geophys. Geosys.* 14:839-850.

Tsakas, S. C. and David, J. R. 1986. Speciation burst hypothesis-an explanation for the variation in rates of phenotypic evolution. *Genet. Sel. Evol.* 18:351-358.

Vashisht, A. and Nagarajan, S. 2008. Exposure of seeds to static magnetic field enhances germination and early growth characteristics in chickpea (*Cicer arietinum* L.). *Bioelectromagnetics* 29:571-578.

Xu, C. X., Wei, S., Lu, Y., Zhang, Y., Chen, C., and Song, T. 2013. Removal of local geomagnetic field affects reproductive growth in *Arabidopsis*. *Bioelectromagnetics* 34:437-442.

Xu, C. X., Yin, X., Wu, C. Z., Zhang, Y. X., and Song, T. 2012. A near-null magnetic field affects cryptochrome-related hypocotyl growth and flowering in *Arabidopsis*. *Adv. Space Res.* 49:834-840.

5

Electricity from Living Plants Myth or Reality?

Garggi G, Sreepriya S* and *Girija T
Department of Plant Physiology, College of Horticulture
Kerala Agricultural University, Thrissur, Kerala

Introduction

Clean and renewable energy is a basic requirement for developmental activities. Global energy crisis is caused by increasing world population, escalating demand due to urbanisation, and continued dependence on fossil-based fuels. Non-renewable energy sources like coal, natural gas oil contribute 79 per cent of world electricity generation; only 21 per cent comes from renewable energy sources. Promoting the use of existing renewable energy sources and identifying new environment-friendly technology is gaining importance. Switching to renewable energy sources like wind energy, ocean energy and solar energy is gaining a lot of importance. These green technologies emit less greenhouse gases into the atmosphere and hence are pollution free and immensely available. Recently, investigators have identified plants as weak energy sources. Plants are versatile organisms capable of communicating with each other with a number of signaling mechanisms. It was observed that plants generate electronic signals inside their system. The concept of utilising electronic signals for studying various physiological plant processes was first put forth by Sir Acharya J. C. Bose, an eminent biophysicist, botanist, archaeologist and polymath from India (Fromm and Lautner, 2007). His ideologies and concepts of movements of plants laid the foundation for modern electrophysiology- a branch of physiology that deals with electrical phenomena associated with metabolic functions. Attempts have been made to harvest electricity from free electrons available in plant leaves produced during photosynthesis and from rhizosphere of plants formed as a result of bacterial activity. Currently, a lot of attention is being given to improvise and develop this technology. To proceed forward in this direction, it is very critical to understand the mechanism of electrical signaling in plants.

Inter-cellular electrical communications in plants

Plants differ from other living organisms due to its sedentary nature. Like all living organisms, plants also respond to stimuli. They receive a variety of signals from the environment, biotic and abiotic cues, which they convert into electronic signals for communication within the system. Various internal signals are required for coordinating expression of genes during development. The process of communication inside the plant system is called signal transduction. Signal transduction is a cascade of events starting from the receiver to the effector molecules, which propagates the information for inter-cellular communication (Taiz and Zeiger, 2010).

Electrical signals in plants was first identified while scientists were trying to explain the mechanism behind the sensitivity of plants like *Mimosa pudica* and Venus trap fly. Physiological and anatomical studies done on *Mimosa pudica* revealed that, propagation of electrical signals was responsible for sensitivity of the plant to 'touch' (Thigmotropism) (Bose, 1925). *Mimosa pudica* contains long slender branches, called petioles, which can fall due to mechanical, thermal, or electrical stimuli. Petioles contain small pinnules, arranged on rachis or midrib of pinna. Pinnules are the smallest leaflets of a leaf while the entire leaf contains the petioles, pinnae, and pinnules. A pulvinus is a joint-like thickening at the base of a plant leaf or leaflet that facilitates thigmonastic movements. Primary, secondary, and tertiary pulvini are responsible for the movement of the petiole, pinna, and leaflets, respectively (Fromm and Lautner, 2007).

Generation of electrical signals in plants is due to the presence of nutrient elements in ionic form inside the plant system. There is a definite compartmentalization of ions inside the plant system among different cell organelles. Differences in this ionic concentration creates a potential difference, which is responsible for their diffusion.

H^+-ATPases generate proton electromotive forces across the plasma membrane and the vacuolar membrane. Anions are actively taken up into the cytoplasm by anion/proton symport systems operating at the plasma membrane. Negative plasma membrane potential together with anion concentration gradient drive passive fluxes of anions out of the cell through plasma membrane anion channels. Moreover, high anion concentration in vacuoles may result from passive anion fluxes driven by negatively charged tonoplast, through vacuolar anion channels.

Major cellular functions of plasma membrane anion channels are to regulate the movement of ions. Due to the highly negative transmembrane potential and outward-directed anion gradients across the plasma membrane, opening of anion channels results in anion release from cytoplasm to the extracellular space. Anion channel activity is coupled with the activity of other transporters, such as Ca^{2+}channels, H^+-ATPase or K^+ channels and contributes to three major functions:

i. Electrical signaling and calcium signaling;
ii. Control of membrane potential and pH gradients
iii. Osmoregulation.

Opening of anion channels results in anion release from cytoplasm to extracellular space. Anion channel activation will thus induce membrane depolarization, which in turn may activate other voltage-dependent channels, such as Ca^{2+} channels, or serve as an electrical shunt for H^+-ATPase which extrudes H^+ ions into the apoplast. These aspects may contribute to short-term electrical and/or calcium signaling, and to the regulation of membrane potential and pH gradient across the plasma membrane (Brygoo *et al.*, 2000).

How are electrical signals perceived by plants?

Signal transduction pathway involves three main steps namely perception, transduction and response. Perception means receiving the stimulus; this is by receptor molecules which are normally located on the cell membrane. When the stimulus binds with the receptor molecules, it circulates in the cell through the process of transduction. During transduction, signals will be amplified by generation of second messengers which carry them to the effector molecule. When the stimulus reaches the effector molecules, a corresponding biological effect is activated with the help of relay molecules. This is the third and final step and is known as "response". It is initiated with an upregulation or down regulation of a corresponding gene expression (Taiz and Zeiger, 2010).

Types of signals generated by plants

There are different forms of signals with which messages are conveyed from cell to cell in plants. Different plant signalling systems are shown in figure 1.

Apart from electrical signals, there are hydraulic and chemical signals in plants for inter cellular communication in plants.

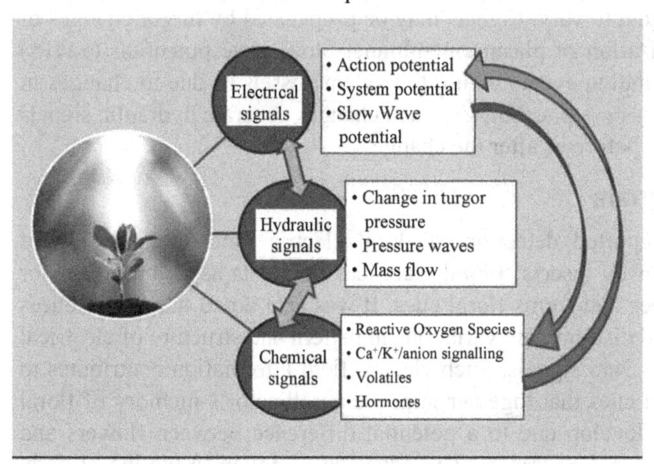

Fig. 1: Plant signaling systems: Types and interactions.
(Adapted from Huber and Bauerle, 2016)

1. Electrical signals propagate as action potential, system potential and slow wave potential.
2. Hydraulic signals include changes in turgor pressure, pressure waves and mass flow of ions.
3. Chemical signals include reactive oxygen species, cation and anion channel signalling, volatile chemicals, hormones *etc.*

All the three signalling pathways are interwoven for effective communication between plants cells (Huber and Bauerle, 2016). Interaction between the different signalling pathways contribute to changes in gene expression pattern at molecular level and result in biological effects.

Systemic signalling

Systemic signalling in plants occurs in response to a wide range of stimuli. When plants are exposed to different stresses in a locality, they perceive many stimuli which then elicits a wider adaptive response in plants. Once the signals are perceived by plants, it triggers a systemic transmission of information. This transmission may be carried out with a range of factors. These factors can be Volatile Organic Compounds (VOCs), small molecules, hormones, RNAs, proteins, reactive oxygen species (ROS), Nitrous Oxide, Ca^{2+} or even hydraulic or electrical signals (Choi *et al.,* 2016). Such signalling mechanisms in plants has been observed to impart Systemic Acquired Resistance (SAR) and Systemic Acquired Acclimation (SAA) in plants. SAR is acquired by plants by exposure to biotic stress whereas SAA results from exposure to abiotic stress (Gilroy *et al.,* 2016).

Types of electrical signalling

Electrical signals include action potentials (APs), slow wave potentials (SWPs), system potentials (SPs), and wound potentials (WPs). Production pathways of these signals and their velocity are found to vary. Signals may be propagated by turgor changes of cells leading to depolarization of plasma membrane. Slow wave potentials (SWPs) may be the result of cavitation events within the xylem vessels or due to changes in turgor such as plasmolysis or deplasmolysis. Such changes generate hydraulic signals in the xylem tissue either before or after the change.

Floral electrical signals

Robert *et al.,* (2013) reported detection of floral electric fields around flowers which require pollination by insects. Floral electrical field acts as an attractant for the naturally charged bees and forms floral cues. It was also noted that bumblebees (*Bombus terrestris*) could discriminate variations in pattern and structure of electrical fields (signals) just like visual signals. Such electric field information contributes to a complex array of floral cues that together improve a pollinator's memory of floral rewards. Electric fields develop due to a potential difference between flowers and insects. This also helps in pollen transfer and adhesion of pollen to the insect body over short distances. Since floral electric fields change within seconds, this sensory modality may facilitate rapid and dynamic communication between flowers and their pollinators. In addition to this, insects use several other senses such as colour, shape, pattern and volatiles for foraging.

Measurement of electrical signals in plants

Electrical signals in plants are detectable as a physical quantity, impulse, voltage or electric currents. Pioneering methods to measure variation in pattern and structure of electrical signals in plants dates back to 1900's, when Sir J. C. Bose detected electrical disturbances in vascular bundles of fern using a galvanometer as early as in 1926. Another scientist, Umrath recorded action potential using micro electrodes in *Nitella* cells in 1930. Propagation of electrical impulses in *Mimosa* was observed by Siboka in 1950. Backster in 1968 found that polygraphs (lie detectors) can be used to measure electrical conductions in plants (Fromm and Lautner, 2007). A number of techniques suggested by Lautner in 2007 for the measurement of electrical signals in plant systems are given below.

Measurement of extra-cellular signals

Extra-cellular electrical signals are recorded using electrodes attached to the portion in between plant parts. It is based on bioelectric activity of plants. Metal electrodes are inserted into plants and a reference electrode is simultaneously inserted into soil for recording extra-cellular measurements.

Measurement of intra-cellular signals

To measure intra-cellular electrical signals, glass electrodes are inserted into cells. This method can be used to measure membrane potentials and electrical currents. Long-distant electrical signals in phloem can be recorded using this method

Aphid technique

In this method, aphids are allowed to settle on leaves overnight. Later, stylet of the aphid is removed by laser pulse, keeping the stylet intact. Then, electrodes are inserted to the stylet and measurements with respect to stimuli are recorded.

Plant electro stimulation

This is a technique to evaluate biologically closed electrical circuits in plants. Cells of many biological organs generate electric potentials that can result in flow of electric currents. Electrical impulses may arise as a result of stimulation. Once initiated, these impulses can activate adjacent excitable cells. Changes in transmembrane potential can create a wave of depolarization which affects the adjoining, resting membranes. When plasma membrane is stimulated at any point, the action potential can propagate over the entire length of cell membrane along the conductive bundles of tissue with constant amplitude, duration, and speed. In electrostimulation, charged capacitor applies electrical potential between two electrodes decreasing gradually to zero during the discharge of a capacitor. A function generator maintains a given high or low potential in the electrodes. Electrical response during potential increase or decrease from a function generator is measured. Once the information is recorded, from the data, it is possible to convert the analogue physical signals into digital numeric values that can be stored, processed, and visualized in a computer (Jovanov and Volkov, 2012). Activation of such electrical circuits can lead to various physiological and biochemical responses.

Patch clamp technique

This is a widely used technique in electrophysiological studies. It was first demonstrated by the German cell physiologist Erwin Neher and Bret Sakmann in 1990. They developed the technique in late 1970s and early 1980s. Their work helped to measure the current of a single ion channel molecule for the first time. This technique improved the understanding of the involvement of channels in fundamental cell processes such as action potentials and nerve activity. They were awarded the Nobel Prize in 1991 in physiology for their contribution. The basic principle in this technique is to isolate a patch of membrane electrically from the solution and record the current flowing into the plant (Yan *et al.,* 2009). Though the technique is widely used in human physiology, it can be successfully utilised to measure changes in membrane voltage also known as membrane potential.

Biomachine

Application of biorobotics has led to the development of a robotic machine called biomachine. Here, the input signals are different light modes, based on which the biomachine will operate. Frequency counters are used to record the response of plant to different light conditions; the frequency is then converted into voltage signal by using the IC, LM2907. Direction changes in the biomachine can be observed by varying light modes which results in change in frequency of action potential signals from plants. These can be detected with an algorithm based software (Aditya *et al.,* 2011).

Sources of electrical energy in living plants

Photosynthesis and rhizodeposition are main sources from which electrical energy can be harvested from living plant. Photosynthesis is the incredible aspect of nature's ability to convert solar energy to chemical energy through a complex series of reactions. Plants excrete organic matter into the soil known as root exudates which helps in the symbiotic association between plants and rhizosphere bacteria. This organic matter is broken down by bacteria living near the root zone. In this breakdown process, electrons released are a waste product, which can be harvested with inert electrodes and converted into electricity, without affecting plant's growth in any way. A recent research from Wageningen University proved that rhizodeposition of plants can be used as a potential substrate for electricity generation.

Electric energy from photosynthesis

In photosynthesis, release of electron occurs during water splitting process in thylakoid membrane during light reaction. Plants function with 100% quantum efficiency, *i.e.*, every photon captured by the plant is converted in to an electron. For a comparison, commercially available solar cells have only 12–17% quantum efficiency (Ramasamy *et al.*, 2013).

Ryu and coworkers in 2010 used a nanoprobe to directly extract photosynthetic electron from an algal cell, *Chlamydomonas reinhardtii.* They inserted a nano-electrode into the chloroplast and obtained a current of 1.2 pA from one cell with a current density of 6000 mA m^{-2}.

Ramasamy and coworkers in 2013 have immobilized the spinach thylakoids onto multiwalled carbon nanotubes (MWNT) using a molecular tethering chemistry. The resulting thylakoid–carbon nanotube composites showed high photo-electrochemical activity under illumination. Multiple membrane proteins have been observed to participate in direct electron transfer with the electrode, resulting in the generation of photocurrents, the first of its kind reported from natural photosynthetic systems. Upon inclusion of a mediator, the photo-activity was enhanced. The major contributor to photocurrent was light-induced water oxidation reaction at the photo system II complex. The thylakoid–MWNT composite electrode yielded a maximum current density of 68 mA cm^{-2} and a steady state current density of 38 mA cm^{-2}, which was larger than previously reported for similar systems. High electrochemical activity of thylakoid–MWNT composites has significant implications for both photosynthetic energy conversion and photo fuel production applications. A fuel cell type photosynthetic electrochemical cell developed using a thylakoid–MWNT composite anode and laccase cathode produced a maximum power density of 5.3 mW cm^{-2}, comparable to that of enzymatic fuel cells.

Scientists in France have transformed chemical energy generated by photosynthesis into electrical energy by developing a novel bio fuel cell. This bio fuel cell functions using the product of photosynthesis (glucose and oxygen) and is made up of two enzyme modified electrodes. Power generated is 9μWcm^{-2} under 250 Wcm^{-2}.

Choo and coworkers in 2013 have harvested energy by embedding electrodes into aloe vera plant to allow flow of ions and hence generate electricity. Multiple random tests were conducted using different types of electrodes as an attempt to determine the characteristics of the harvesting system. It was found that voltage is produced to a greater or lesser extent in all the tests where combination of copper-zinc electrodes was used.

Apart from these Low *et al.,* (2013) observed that in spinach leaves, chlorophyll samples within the range of concentration from 76.46 mg ml^{-1} to 88 mg ml^{-1} could light up the LED lamps which indicated that chlorophyll from plants can be a new source of alternative energy.

Electricity from rhizodeposits

Plants capture light energy during photosynthesis. In this process, carbon dioxide and water is taken up and converted into chemical bonds of sugars. Part of this chemically stored energy is transferred via the roots and littered into the soil called rhizodeposits. This energy transported into the soil can be captured by the so-called electro-chemical active bacteria. These micro-organisms are capable of oxidizing the organic matter and transferring the energy rich electrons to an electrode. The energy carried by the electrons can be used as electrical energy, after which the electrons react at another electrode with oxygen to form water (Timmers, 2012). This technology is called the Plant-Microbial Fuel Cell (Plant-MFC).

Proof of principle of plant microbial fuel cell

Reed manna grass (*Glyceria maxima*) is a common grass species in America, Europe, and Asia. The principle of plant microbial fuel cell was tested in this species by placing its roots in the bio-anode compartment of the microbial fuel cell. A cell voltage of 6 was generated with one or two plants and without any plants only a cell voltage of two was recorded during a period of 118 days. An incubation period of 50 days proved to be necessary to obtain conditions favorable for electricity generation. The cell voltage of both plant-MFCs increased steadily from day 50 and reached maximum of 253mV in the fuel cell with one or two plants by the 72nd day and, 217mV in the second cell with no plants by the 66th day. This corresponded to a current generation of 0.253 and 0.217 mA respectively.

Rhizosphere-anode model

In 2010, Strike and co-workers prepared a Rhizosphere-anode model with roots as a source of exudates and oxygen. In this model, oxygen loss and exudation were related to microbial growth. Living roots release exudates (electron donors) and oxygen (electron acceptor). Normally, in the presence of oxygen, aerobic bacteria will out-compete electron active bacteria (EAB), since oxygen is the preferred electron acceptor over the graphite electrode. Oxygen active bacteria (OAB) will develop adjacent to the root. Here the aerobic biofilm is the aerobic bacteria. EAB will develop only when oxygen is depleted by the aerobic biofilm and exudates are available. The aerobic biofilm has three major functions which decides the (i) the mass balance of oxygen, (ii) the mass balance of the electron donor, and (iii) the mass balance of the biomass. The amount of exudates available for the EAB is now equal to the flux of exudates out of the aerobic biofilm per mol $cm^{-2}s^{-1}$. Acetate is chosen as a model exudate because it is a suitable substrate for EAB and is a major constituent of exudation. The model cell will contain both an oxic zone and an anoxic zone. The current generated in the anoxic zone will be harvested.

Electricity from rice fields

Wetlands in India account for 18.4% of country's geographical area, of which 70% is under paddy cultivation. The technology of Plant-Microbial Fuel Cell (PMFC) can be successfully integrated in paddy fields. The presence of anaerobic condition, presence of rich organics, availability of root exudates and the existence of a potential gradient between soil and flooded water in rice fields are the factors suitable for generating electricity by PMFC from rice fields.

Kaku and co-workers in 2008 have successfully integrated PMFCs in rice fields and obtained a power output of 6 mWm^{-2}. In another study done by Takanezawa *et al.*, (2010), PMFC systems were set under different experimental conditions and power outputs from these PMFC was evaluated. The study showed that organic exudates from roots could be utilized for microbial anode respiration, if the roots were contacted with more anodes. The anode position, both depth of anode and distance between the anode and cathode, influenced the performance of PMFC as it affects the proton diffusion from anode to cathode. Placing anode at a depth of 5 cm was better than placing them at 2 cm; this may be because 2 cm region may not be anaerobic and presence of oxygen in this area may serve as an alternate electron sink. When anode is placed at 5 cm depth, more plant roots may get access to the anode, resulting in larger amount of organic supply for the roots.

Bombelli and coworkers (2012) have successfully compared voltage output from different plant species. The plants used for the study were: C3 plant *Oryza sativa* and C4 plant *Echinochloa glabrescens*. Maximum power output generated by *E. glabrescens* was almost 10 times lower than the maximum power output from *Oryza sativa* and both showed circadian oscillation in electricity generation.

Microbial communities at the anode PMFCs were analyzed to elucidate principles and performance of PMFCs. It was observed that the most common bacteria were from the families *Desulfobulbus* or *Geobacteraceae* (De Schamphelaire *et al.*, 2010) or were closely related to *Natronocella, Beijerinckiaceae, Rhizobiales* and *Rhodobacter* (Kaku *et al.*, 2008). Some species, like *Geobacter sulfurreducens*, were also found to be electrochemically active (Bond and Lovley, 2003). The microbial organisms that were suitable for production of PMFC in *Glyceria maxima* were identified as members of the family *Geobacteraceae, Clostridiaceae, Ruminococcaceae* and *Comamonadaceae*. The electricity generation was mainly due to the presence of *G. sulfurreducens* and *G. metallireducens* (Timmers *et al.*, 2012). The performance efficiency of different PMFCs is given in Table 1.

Table 1: Performance and efficiency of plant microbial fuel cells

Plant	Microbial community	Operation time (days)	Current density (m A/m^2)		Power density (%)
			Avg.	Max.	Max.
Oryza sativa ssp. *indica*	*Desulfobulbus* sp. *Geobacteraceae*	134	44	_	33
Oryza sativa L.cv. *sasanishiki*	*Natronocella* *Beijerinckiaceae* *Rhizobiales*	120	–	52	6
Glyceria maxima	Bacteria	67	32	153	67
Spartina anglica	Bacteria	78	141	_	79
Arundinella anomala	Bacteria	112	–	–	22

Electricity generation potential of PMFC

Climate change mitigation requires a global change in the perspective of power generation. In this context, a multidisciplinary European research consortium is working towards optimal electricity production of 1000 GJ ha^{-1} year^{-1} (3.2 Wm^{-2}) from plant resources. A first estimate for electricity production by a PMFC under Western European conditions was 21 GJ ha^{-1} year^{-1} (67 mWm^{-2}). They tried to develop a realistic estimate after computing the average values of natural resources in the Western European countries of Netherlands, Belgium and France. This was computed based on the estimates available on the (i) average solar radiation of 150 Wm^{-2}; (ii) average photosynthetic efficiency of 2.5%; (iii) common rhizodeposition of 40%; (iv) rhizo deposit availability for microorganisms of 30%; (v) MFC energy recovery of 29% including a growth season of 6 months in 2010. A PMFC of *Spartina angalica* yielded a power output of 50m Wm^{-2}.

Electricity producing green roofs

The latest application of PMFC is its use as green roof. This is becoming increasingly popular especially in cities, as these green roofs will help to improve air quality, serves as an insulation for buildings and reduces urban heat. It also improves aesthetic value of buildings, reduces the impact of storm and water runoff which in turn will increase biodiversity. Applying PMFC to green roofs combines the advantages of these roofs with electricity generation. Studies in Netherlands show that a reasonably sized roof of 50 m^2, on a flat roof surface could produce 150W continuously when the proposed maximum of 3.2 Wm^{-2} is reached. Assuming an average electricity need of 500W, a green roof could provide approximately one-third of a household's electricity requirement. Moreover, energy use by the household will decrease owing to the insulation capacity of the green roof, so PMFC power could be expected to account for a larger proportion of the household's energy need. This technique is being tried all over the world.

Trade-off between PMFCs, wind turbines and solar panels

As compared to wind turbines, photovoltaic solar panels and other alternative renewable electricity sources, PMFCs are more desirable as they improve aesthetic value of the landscape and also protect biodiversity. However, a cost-benefit analysis is imperative when a new technology for electricity production is being implemented.

In a natural environment, maximum estimated power yield from PMFC is only 1.6 M Wkm^{-2}. Whereas wind turbines could generate 5–7.7 MW km^{-2} on a typical wind farm in Europe, solar panels could generate 4.5–7.5 MW km^{-2} under Western European conditions (solar radiation 150 Wm^{-2}, PCE 15–25%) the tilted position of a solar panel thus uses 2.5 m^2 of land per m^2 of solar panel. Power output of wind farms and solar farms were found to be three- to five-fold higher than that of PMFCs. There is an increasing need for electricity and light from renewable resources with least impact to the environment and biodiversity. Both wind turbines and solar panels seem to have a lot of disadvantages wherein, while turbines contribute to avian mortality, noise and electromagnetic interference, solar panels have polluting metals, need space leading to loss of green space and biodiversity and increasing dark surface. PMFCs could offer an opportunity for electricity generation while sustaining the natural environment at locations where wind turbines or solar panels are not desirable. Future integration of PMFCs into closed systems could provide 24 h/day electricity generation without the use of scarce materials and with nutrient preservation. Recent discoveries such as potential of cuticle-cellular tissue bilayer in higher plant leaves to function as an integrated electric generator, capable of converting mechanical stimuli in to electricity can help to improve the efficiency and commercial viability of this technology.

e- Plant: the concept

Application of biophysical tools in botany leads to the concept of e-plant which forms the basics of electrophysiology. According to Stavrinidou *et al.,* (2015) e-plant utilizes electronic technology to study natural vascular circuitry system of plant. Here, xylem, leaves, veins and signals of plants are utilised as components of an electronic circuit to record and regulate plant metabolism. Plants are live machines and have electro-conductive ability. This is utilised in developing e- plants.

A team of scientists from the Organic Electronic Laboratory, Linkoping University, Sweden created the first electronic plant. With the help of channels that distribute water and nutrients in plants, the team succeeded in building the key components of electronic circuits. The model plant they used was, rose (*Rosa floribunda*). They showed that rose can produce both analog and digital electronic circuits which can be further exploited to regulate plants' physiology. Electrical signals generated in plants are transported over long distances through xylem and phloem vascular circuits, where they are triggered, modulated and power processed throughout the plant system. This vascular circuitry can be regulated artificially. An electronic technology to leverage plants' native vascular circuitry to harvest new pathways and other biochemical process are aimed in the development of electronic plants.

According to Harpreet Sareen (2019) of Massachusetts Institute of Technology, plants are active signal networks that are self-powered, self-fabricating, and self-regenerating systems at scale. Both Harpreet Sareen and Pattie Maes have contributed to "Cyborg Botany" which combines the field of plant electrophysiology and plant neurobiology to develop cyborg plants.

The discovery of a conducting polymer named poly (3,4-ethylenedioxythiophene) (PEDOT) was the breakthrough event in creating the electronic plant. This polymer has the property of electronic and ionic conductivity in hydrated state. PEDOT was doped with polystyrene sulfonate (PEDOT:PSS) to make it biocompatible. With this, they integrated analog and digital circuits and logic gates. Internal structures of the plant were used as a template for *in situ* fabrication of electronic circuits.

Scientists used two methodologies, namely excised shoot tip and leaf vacuum infiltration. For excised shoot method, the lower part of garden rose was cut and the fresh cross section was immersed in an aqueous PEDOT-S:H for 24- 48 hours. During this time the solution was taken up into the xylem vascular channel and transported apically. After taking out rose from water and rinsing, the outer bark was peeled off, which exposed the dark continuous lines of the impregnated polymer wire. Scanning electron microscopic images clearly visualised the polymer filled xylem. For leaf vacuum infiltration, PEDOT:PSS combined with nanofibrillar cellulose (PEDOT:PSS-NFC) was used.

The key findings in the development of e-Plant were that scientists successfully integrated organic electrical circuits *in vivo,* which can be utilised to measure the electrical conductivity in plants along with induction of electrochromism in leaf, suggesting that external manipulation of internal functions is possible with variation in externally given potential difference and this can be exploited to identify the ion migration pathway in leaf. Thus e-Plant serves as an electronic tool to manipulate physiological processes.

Significance and utilisation of e-plant concept

e-Plants can be an excellent tool to measure the concentration of ionic molecules in plant parts. The external aided control and interaction with metabolic biochemical pathways can be materialised using e-Plants. Thus, the internal functions of plants can be modulated at our will. It can also be utilised as a technique to exploit the excess energy lost from the plant. Incorporation of suitable sensors in circuits can help in building power plants. Development of e-plants is highly significant, since it integrates plant biology and electronics, which is an unusual combination. Using this methodology, electronic signals required to initiate any biological process can be delineated, with which metabolic regulation can be done. e-Plant can serve as a complimentary methodology for genetic manipulation techniques. Moreover, it is a novel way to communicate with plants. Thus plants can be programmable computing circuits in future.

Plant-e Company

"Plant-e" is a Dutch company that is focused on developing products in which electricity is generated with living plants. This very innovative method of electricity production is not only very friendly for the environment; it is also unique and can be widely integrated around the world. Plant-e was founded on September 14, 2009 as a spin-off company from the sub-department of environmental technology of Wageningen University. In 2007, the technology was patented. Plant-e's products are currently built up as modular system. Plant-e is developed as a tube system that is placed beneath the surface of the ground, which can be applied in existing wetlands. The plant-e company harnesses electricity from living plants, and then uses it to power cell phone chargers, Wi-Fi hotspots, and now over 300 LED streetlights in two sites in the Netherlands. Plant power is also being used near the company's headquarters in Wageningen. In November 2014, Plant-e launched its starry sky project at an old ammunition site called Hamburg, near Amsterdam.

Voltree power is another American company dedicated to research in production of electricity from trees without destroying the environment.

Conclusion

Living plants are sustainable and environment friendly sources of electrical energy. Photosynthesis and rhizodeposition can be exploited to generate electricity by understanding the intricacies of electrical signaling in plants. Electrical energy from living plants can be complementary to conventional bio energy systems. Plant microbial fuel cell is a very innovative technology developed in the Netherlands which can be suitably integrated in wetlands. This technology is economically feasible for remote areas where there is a lack of infrastructure for storage of electrical energy. The concepts of e-Plant and their integration into day-to-day activities can be realized with sufficient research and investment partners.

References

Aditya, K., Udupa, G., and Lee, Y. 2011. Development of bio-machine based on the plant response to external stimuli. *J. Robot.* 10: 1-7.

Bombelli, P., Iyer, D., Susan, M. R., Harrison., T. L., and Christopher, H. 2012. Generation of current in vascular plant bio-photovoltaic systems based on rice (*Oryza sativa*) and an associated weed (*Echinochloa glabrescens*). *Appl. Microbiol. Biotechnol.* 10: 4473-4476.

Bond, D. R. and Lovley, D. R. 2003. Electricity Production by *Geobacter sulfurreducens* Attached to Electrodes. *Applied and Environmental Microbiology* 69 (3): 1548-1555.

Bose, J. C. H. 1925. Physiological and anatomical investigations on *Mimosa pudica. Proc. Royal Soc.* B98:280-99.

Brygoo, B. H., Vinauger, M., Colcombet, J., Ephritikhine, G., Frachisse, M. J., and Maurel, C. 2000. Anion channels in higher plants: functional characterization, molecular structure and physiological role. *Biochem. Biophys. Acta* 1465: 199-218.

Choi, W.G., Hilleary, R. Swanson, S. J. Kim, S. W., and Gilroy, W. 2016. Rapid, long-distance electrical and calcium signaling in plants.*Annu. Rev. Plant Biol.* 67:287–307.

Choo, Y. Y. and Dayou, J. 2013. A Method to Harvest Electrical Energy from Living Plants. *Journal of Science and Technology* 5(1).

De Schamphelaire, L., Cabezas, A., Marzorati, M., Friedrich, M. W., Boon, N., & Verstraete, W. 2010. Microbial community analysis of anodes from sediment microbial fuel cells powered by rhizodeposits of living rice plants. *Applied and Environmental Microbiology* 76(6): 2002–2008.

Flexer, V. and Mano, N. 2010. From dynamic measurements of photosynthesis in a living plant to sunlight transformation into electricity. *Anal. Chem.* 82(4): 1444-1449.

Fromm, J. and Lautner, S. 2007. Electrical signals and their physiological significance. *Plant Cell Environ.* 30: 249-257.

Gilroy, S., Białasek, M., Suzuki, N., Górecka, M., Devireddy, A. M., Karpiński, S., and Mittler, R. 2016. ROS, calcium and electric Signals: key mediators of rapid systemic signaling in plants. *Plant Physiol.* 1104: 1-16.

Huber, E. A. and Bauerle, L. T. 2016. Long-distance plant signaling pathways in response to multiple stressors: the gap in knowledge. *J. Exp. Bot.* 1093: 1-17.

Jovanov, E. and Volkov, A. G. 2012.Plant electrostimulation and data acquisition. In: Volkov, A. G. (ed) *Plant Electrophysiology*. Springer, Berlin. pp. 45-67.

Kaku, N., Yonezawa, N., Kodama, Y., and Watanabe, K. 2008. Plant/microbe cooperation for electricity generation in a rice paddy field. *Appl. Microbiol. Biotechnol.* 79: 43-49.

Low, H. B., Dayou, J., and Wong, N. K. 2013. Chloropyll as a new alternative energy source. *Int. J. Sci. Envt.* 2(3): 320-327.

Ramasamy, R. P., Calkins, J. O., Umasankar, Y., and Neill, H. O. 2013. High photo-electrochemical activity of thylakoid-carbon nanotube composites for photosynthetic energy conversion. *Energy and Environ. Sci.* 6: 1891-1900.

Robert, D., Clarke, D., Whitney, H., and Sutton, G. 2013. Detection and learning of floral electric fields by bumblebees. *Sciencexpress* 10 (1126): 1-7.

Ryu, W., Bai, S., Huang, Z., Moseley, J., Fabian, T., Fasching, R. J., Grossman, A. R., and Prinz, F. B. 2010. Direct extraction of photosynthetic electrons from single algal cells by nanoprobing system. *Nanoletters*10(4): 1137-1143.

Stavrinidou, E., Gabrielsson, R., Gomez, E., Crispin, X., Nilsson, O., Simon, D. T., and Bregrren, M. 2015. Electronic plants. *Sci. Adv.* 1:1-8.

Strik, D. P. B. T. B., Hamelers, H. V. M., Snel, J. F. H., and Buisman, C. J. N. 2008. Green electricity production with living plants and bacteria in a fuel cell. *Int. J. Energy Res.* 32: 870-876.

Strik, D. P. B. T. B., Timmers, R. A., Heldel, M., Steinbusch, K. J. J., Hamelers, H. V. M., and Buisman, C. J. N. 2011. Microbial solar cells: applying photosynthetic and electrochemically active organisms. *Trends in Biotechnol.* 29(1): 41-49.

Strik, D. P.B.T. B., Hamelers, H.V. M., and Buisman, C.J.N. 2010. Solar energy powered microbial fuel cell with a reversible bioelectrode. *Environ. Sci. Technol.* 44 (1): 532-537.

Taiz, L. and Zeiger, E. 2010.*Plant Physiology* (6th Ed.), Sinauer Associates. Massachusetts, 690p.

Takanezawa, K., Nishio, K., Kato, S., Hashimito, K., and Watanabe, K. 2010. Factors affecting electric output from rice-paddy microbial fuel cells. *Bioscience, Biotechnology, and Biochemistry* 74 (6):1271–73.

Timmers, R. 2012. Electricity generation by living plants in a plant microbial fuel cell. Ph. D thesis, The Wageningen University, Netherlands, 220p.

Timmers, R., Rothballer, M., Strik, D. P. B. T. B., Engel, M., Schulz, S., Schloter, M., Hartmann, A., Hamelers, B., and Buisman, C. 2012. Microbial community structure elucidates performance of *Glyceria maxima* plant microbial fuel cell.*Appl. Microbiol. Biotechnol.* 94: 537–548.

Yan, X., Wang, Z., Huang, L., Wang, C., Hou, R., Xu, Z., and Qiao, Z. 2009. Research progress on electrical signals in higher plants. *Prog. Nat. Sci.* 19: 531-541.

6

Plant Architecture: Evolution Diversity, Regulation and Scope

Nithya N, Amrutha E A* and *Girija T
Department of Plant Physiology, College of Horticulture
Kerala Agricultural University, Vellanikkara, Thrissur, Kerala

Introduction

Plant architecture is defined as a three-dimensional organization of plant body in terms of its size, shape position of leaves, flower organs and branching pattern. Earlier this was the only criterion for systematic and taxonomic classification of plant species and the best means of identifying them (Stecconi, 2006). Study of plant architecture emerged as a new scientific discipline some 30 years ago. The subject derives its base from earlier works on plant morphology (Halle and Oldeman, 1970) but currently it is integrated with several disciplines of plant sciences ecology and engineering. Modification in plant architecture causes alteration in primary and secondary growth. This occurs due to differences in phyllotactic arrangement, branching pattern and floral differentiation. Plant architecture can be modulated by various parameters like climate, agronomic practices and human interventions. Manipulation of plant architecture is a major area in plant science which is currently adopted for productivity improvement by breeders, agronomists, horticulturists and landscape managers. However, the basis for such manipulation lies in the knowledge of the basic architecture of plants and its significance.

Evolution of plant architecture

Zimmermann proposed 'telome theory' to describe plant morphogenesis. It also explains major stages in evolution of vascular plants. He was able to visualize the transformation of stem into mesophyll leaf by modification of existing organs and also interpret and explain evolution of diverse structural morphologies of leaves, sporophylls, stems and roots of vascular plants (Berling and Fleming, 2007). His theory is still valid due to its applicability across a wide variety of organs and taxa. According to his theory, all vascular plants developed from "telomes" which are single-nerved terminal units at base or apex. Units or connecting axis below dichotomous branching were called "mesomes", similar to the internodes of present-day plants (Sussex and Kerk, 2001). Evolution of diverse structures is brought about by combination of five elementary process such as:

1. **Overlapping:** This is due to unequal development of two products of a terminally branched apex. It explains how the apical meristem outgrows or overgrows the

other and how the larger axis becomes the stem and shorter overtopping branches represent lateral branches or leaves.

2. **Reduction:** length of telomes and mesomes are reduced leading to development of microphyllous leaves or needle like leaves of conifers.
3. **Planation:** produces telomes and mesomes in a single plane which may be overlapping, it shifts from a three-dimensional pattern to a single plane. leading to the formation of planated branches. Infilling between the planated branches is called webbing. This leads to the formation of flattened leaf like structures or lamina.
4. **Syngenesis**: lateral fusion of sterile telomes and mesomes in a single plane leads to the formation of complex vascular systems and veins of leaves.
5. **Curvation:** In this process reproductive structures such as sporangia are formed when fertile telomes curves out of the plane of branching. It can either be recurvation when the telome bends inside towards an axis or incurvation when the telome bends downwards shifting sporangia from terminal to ventral position.

This theory gave a comprehensive explanation of origin and early diversification of plant architecture (Sussex and Kerk, 2001).

Architectural model

Architectural model is a physical representation of a structure. In plants, it explains the inherent growth strategy that defines both the manner in which a plant elaborates its form and the resulting architecture. In nature, 23 architectural models have been identified and classified. Each of these models named after a well-known botanist is actually a combination of simple morphological features such as orientation of leaves and branching pattern on the main stem and main branches, also morphological differentiation of axes and position of sexual organs. Specific architecture of a plant is the result of fundamental growth programme which establishes the entire architecture. The models help to explain the nature and sequence of activity of the endogenous morphogenetic processes of the organism on which the entire plant architecture is established.

Plant identification based on architectural model is actually based on 4 major simple morphological features of plants. These include

1. **Growth pattern and cycle of growth:** This can be either determinate vs. indeterminate growth or rhythmic vs. continuous growth;
2. **Mode of trunk construction:** This can be monopodial or sympodial. In monopodial, the trunk is formed by the uninterrupted growth of an indeterminate apical meristem while in sympodial, a trunk or branch is formed by the relay growth of a sequence of determinate lateral meristems.
3. **Morphological differentiation of axes and branching pattern:** Plants can have an orthotropic or plagiotropic or even a mixed morphological geometry. In orthotropic type, branches are arranged in ascending axes (vertical orientation) and leaves are borne in radial symmetry while in plagiotropic branching,

arrangement is in horizontal orientation and leaves are arranged in one plane. In mixed morphology a shoot may first grow vertically and then bend over to become horizontal and leaf arrangement will also vary accordingly.

4. **Position of flower:** Either lateral or terminal flowers

Architectural models have been developed by different scientists based on the above given features. Corner's model is concerned with unbranched plants with lateral inflorescences. Leeuwenberg's model consists of a sympodial succession of equivalent sympodial units, each of which is orthotropic and determinate in its growth; and Rauh's model is meant for woody plants where growth and branching are rhythmic, all axes are monopodial and sexuality is lateral. These models seem to have universal applicability to both trees and herbaceous plants and also to plants belonging to tropical as well as temperate regions.

They are defined by few and simple morphological features; it gives only an idea of the elementary developmental pattern of a species. Architectural analysis shows that frequently exhibited morphological features of plants are related to two or three models these intermediate forms show that there is not much variance between the models. It also indicates that all architectures are theoretically possible and that there could be a gradual transition from one to another. Models represent the forms that are most stable and most frequent in nature. Since growth of plants is genetically determined, there cannot be much variation in features, however under extreme ecological conditions their expression is affected by environment. Models also explain the evolutionary pattern or phylogenetic or taxonomic distribution of plants. Moreover, they unravel the ecological importance of these patterns.

Diversity of plant architecture

Plant morphology has been considered as the 'inspiring soul' of plant architecture studies. Diversity of plant architecture results from the differences in plant morphological traits categorized as (1) Phyllotaxis, (2) Branching pattern, (3) Floral differentiation of axes and (4) Root architecture (Barthelemy and Caraglio, 2007).

A. Phyllotaxy

Phyllotaxy is the arrangement of leaves on the stem. In vegetative phase plants continuously form new leaves that are arranged in regular patterns (phyllotaxis) with defined divergence angles between successive leaves. Common phyllotactic patterns are distichous, decussate and spiral. In distichous phyllotaxis, leaf divergence angle will be 180° and leaf arrangement will be alternate in two opposite rows (Eg. *Zea mays*), whereas, in decussate plants leaf divergence angle will be 90° (Eg. *Mentha piperita)*. Spiral arrangement is seen in plants like *Aloe polyphylla* with a divergence angle of 137° (Reinhardt and Kuhlemeier, 2001).

Regulation of phyllotaxy

To achieve defined divergence angles between successive leaves, the shoot apical meristem (SAM) must integrate spatial information from pre-existing leaf primordia. To explain phyllotactic patterning, "inhibitory field hypothesis" suggested by Schoute

(1913) is currently accepted. This hypothesis proposes that each formed organ primordium generates an inhibitory field which can be chemical or mechanical signals that prevents initiation of new primordia in its vicinity (Douady and Couder, 1996; Hofmeister, 1868; Snow and Snow, 1962). Therefore, as the SAM grows, incipient primordia can be initiated at sites where inhibition is the lowest. Depending on its range and stability, such an inhibitor could create a field that constrains the formation of new leaves to position with defined minimal distance.

Apart from this, plant hormones also play a crucial role (Benjamins *et al.*, 2001). Among the growth regulators, auxin has been firmly established as a central regulator of phyllotaxis. Auxin distribution in tissues have been found to reflect and predict the patterns of organ initiation (Benková *et al.*, 2003; Reinhardt *et al.*, 2003; Heisler *et al.*, 2005; Bayer *et al.*, 2009). Auxin is required for laminar growth so the first indication of organ initiation is the formation of auxin maxima in the epidermis of the meristem, which acts as an inducer of organ formation, and the postulated inhibitory fields around pre-existing primordia reflect low auxin concentrations in their vicinity. It is thought that the two youngest primordia drain auxin from the meristem and thereby determine the position of the next leaf primordium. Leaf angle is also a hormonally regulated architectural trait which is determined by cell size of collar tissue, also called the lamina joint (Cao and Chen, 1995; Zhang *et al.*, 2009; Zhao *et al.*, 2010).

Three modes of auxin movement have been identified in the establishment of different phyllotactic leaf arrangement.

(a) *Distichous Phyllotaxis*: Here the youngest leaf primordia absorb auxin from incipient leaf primordial, where concentration of auxin will be more, resulting in a defined divergence angle;
(b) *Spiral Phyllotaxis*: This type of arrangement occurs when two leaf primordia absorb auxin in a pattern that one of them absorbs to a lesser extent.
(c) *Decussate Phyllotaxis:* Pairs of opposite leaves are formed when the size of the meristem allows two auxin maxima to coexist,

Genes for phyllotaxy

Phyllotaxis is thought to be a multigenic trait as many genetic screens have been identified with regulatory function in leaf formation and positioning. However, they have mostly failed to yield mutants with specific phyllotactic phenotype. Available information indicate that three families of auxin transporters control the distribution of auxin in various tissues of the plant body. Among these, AUX1/LAX proteins are influx carriers (Swarup and Peret, 2012), whereas cellular auxin efflux is mediated by two distinct groups – the PIN proteins (Krecek *et al.*, 2009) and a subgroup of ABC transporters (Kang *et al.*, 2011). Vascular development in plants is controlled by *monopteros* (mp) gene. *KNOX* genes play a major role in meristem maintenance as well as proper pattering in organ initiation (Benjamins *et al.*, 2001). There were concerted efforts in elucidating the mechanism of branch angle characteristics which resulted in the identification of a distinct peach mutant with acute branch angle possessing the causative gene, Tiller Angle Controll (TAC1) (Dardick *et al.*, 2013). TAC1 belongs to a small family of genes, which can increase (TAC1) or decrease (LAZY1) branch angles (Yu *et al.*, 2007; Hollender and Dardick, 2015).

B. Branching

Architecture of some vascular plants consists only of a single vegetative axis during their whole life span, most display a more complex architecture consisting of several axes, one derived from another by a repetitive process known as branching (Halle and Oldeman 1970). Most typical type of branching patterns include monopodial versus sympodial as well as acrotonic versus basitonic branching. Branching is controlled both by genes and also by environment.

(i) **Monopodial Vs Sympodial type:** In monopodial branching, plant continues its growth whereas in sympodial branching growth ceases when it terminates into flowers or some other plant part. Indeterminate meristem typically produces monopodial growth characterized by a pronounced primary stem whereas plants with determinate meristem show sympodial growth. It is a process of repeated loss of SAM (Shoot apical meristem) through terminal differentiation and lateral outgrowth from axillary meristem resulting in compound shoot architecture. Both monopodial and sympodial growth is as a result of differences in the expression of genes and localization of proteins (Busov, 2018).

(ii) **Acrotonic Vs Basitonic:** Acrotony is the general development of lateral axes in the distal part of a parent axis or shoot. Basitony was at first considered as a preferential development of lateral axes in the basal part of a vertical stem (Caraglio and Barthelemy, 2007). Basitony type of branching is seen in bamboo and many grass species.

Physiology of branching

High cytokinin (CK): Auxin (IAA) ratios were observed in rapidly growing shoots at the basal and upper nodes, whereas the ratio was low in the slow-growing middle node shoots. Initially, ABA concentrations did not correlate with rate of shoot growth but later exhibited a strong negative correlation, when CK:IAA ratios did not correlate with rate of axillary shoot growth (Davies *et al.,* 1966). Thus, the bud outgrowth potential is a functional balance of several hormones which varies during development phases. Hence, a focus on regulating hormonal status of particular organs of the plant will help in modifying axillary shoot growth and in turn plant morphology (Huang *et al.*, 2017).

C. Floral Differentiation

Flowering affects plant architecture in many ways. Determinate and indeterminate are the two types of flowering patterns normally observed in plants. Determinate growth is finite. It usually means that the main stem ends with a flower or other reproductive structure while indeterminate growth means sequential flowering that starts at the bottom and on the sides of a plant, and then moves in and up. In many plants, SAM of the main shoot is indeterminate, *i.e* it is active during the entire life span of the plant, producing first leaves and later flowers. This growth behaviour is referred as monopodial (Pineiro and Coupland, 1998). In contrast, SAM of some plants especially

Solanaceae family (eg: tomato) is determinate, *i.e.* it terminates in a single flower and development continues from lateral meristems. This growth behavior is referred as sympodial growth (Coen and Romero, 1994).

Genes involved in floral differentiation

Onset of flowering affects plant architecture in many ways by the activity of different classes of genes. The genes *FLORICAULA (FLO)* and *LEAFY (LFY)* transform indeterminate axillary meristems into determinate floral meristems. Whereas *CENTRORADIALIS (CEN)* and *TERMINAL FLOWER (TFL)* prevent termination and flower formation in the main meristem (Amaya *et al.*, 1999). Genetic analysis of branching in *Solanaceae* has identified genes that regulate sympodial development. In tomato mutant, *self-pruning* (sp) sympodial units are reduced successively.

D. Root Architecture

Major function of plant root is anchorage and resource accumulation. Root system is composed of diverse root types. Roots differ in origin, anatomy and function primary roots are the first to emerge and they are of embryonic origin. Seminal roots are from extra embryonic root primordia. Adventitious roots are from any non-root tissue these can be junction roots which are seen at the root–shoot junction, crown roots which develop from nodes below ground, brace roots which develop from nodes above ground, stem roots that arise from internodes, and hypocotyl roots (Steffens and Rasmussen, 2016). All these root types, including primary roots, can develop into lateral roots. Number, length and growth angle of these constitute root architecture. It decides their distribution in soil and effectiveness.

Root system architecture is an important upcoming area of plant research as architecture and root function can be manipulated to increase resource capture. However, study of root architecture is complex and this should take into account phenology of crop growth and water availability. Generally, by root architecture we mean spatial configuration of the root system and geometric deployment of root axes. According to Benfey and Scheres, (2000) studies of root architecture do not include fine structural details, such as root hairs, but are concerned with an entire root system or a large subset of root system of an individual plant

Developmental processes contributing to root system architecture

Root formation in plant species follow a similar basic architecture. In seed propagation, the first formed primary roots will generally be perennial. This can be adventitious, radical or seminal roots depending on the plant species and this root determines the shape of the root system. Lateral roots will live longer or even become perennial when they undergo secondary growth or when they develop periderm, which can be recognized by the brown colour of the roots. While in fine roots, there will be no secondary growth or wall thickening and these roots will be short lived. This can happen even when the C:N ratio of the soil is low. Root tip diameter is often considered as a morphological marker as it correlates with root longevity.

Events in root architecture development can be classified into 5 steps:

I. Development of primary or adventitious roots is highly coordinated with shoot system growth such as leaf and tiller development.

II. Multiplication of roots by branching, or development of lateral roots, Root elongation and branching are mainly influenced by environmental stimuli especially nutrient and water status of soil. Normally equilibrium is maintained between root number and length since branching and elongation are coordinated processes. When there is a homogeneous condition, in an acropetal branching sequence, lateral roots emergence after a defined thermal time (Heat Units), from the unbranched apical axis, the rate of growth of these lateral branches varies. However, all lateral primordia do not develop to lateral root it is mostly influenced by hormones (Lynch, 1995). Based on nutrient perception, there can be rapid activation of primordia and this promotes growth of lateral roots. This has been verified in *Arabidopsis,* where the effects of N supply on root development indicated that four different aspects were affected by the nutrient.

   1. There was localized stimulatory effect of external nitrate on lateral root elongation;
   2. A high nitrate concentration in the tissues had an inhibitory effect on lateral root meristems activation;
   3. A high C:N ratio suppressed initiation of lateral roots; and
   4. External L-glutamate (organic N) inhibited primary root growth and stimulation of root branching.

III. **Axial growth:** Axial root growth occurs from the distal end (root tip) as a result of both cell division in the meristem and elongation within the elongation zone. Trajectory of the root is influenced by different tropisms and root cap seems to be involved in sensing these. Influence of tropisms varies among root branching order or adventitious/seminal roots.

IV. **Radial growth:** In dicotyledonous plants, radial roots grow laterally from secondary meristems (cambium). These roots show morphological variations and have a range of functions such as increasing axial transport particularly axial hydraulic conductivity, increasing mechanical strength and anchorage, storage and protection against predation, drought and pathogens.

V. **Root senescence and decay:** In many soils roots become defunct after a short period. This is especially true for iron rich conditions, where in the oxidized iron forms a crust on the root surface affecting absorption of nutrients and water. Plants are able to replace decaying roots by more efficient new ones. This is important for resource allocation and improving efficiency of resource accumulation.

Root system architecture diversification

Plasticity of root architecture with respect to its function has been noticed in many crops. Primary and secondary roots are mainly for anchorage while lateral roots and adventitious roots help in foraging nutrients and water from different soil layers.

(i) Root system architecture for anchorage

Anchorage is the primary function of roots. Trees differ in their capacity to give a strong anchorage. Those plants which have major root branches nearest to the stem base and wider root spread angle show greater resistance to lodging.

(ii) Root system architecture for efficient water acquisition

Water is normally found in the deeper soil layers mainly during drought situation. It has been noticed that plants produce root hairs and more adventitious roots in drought tolerant and resistant species. These varieties grow extensive roots for deeper penetration and also partition more nutrients for root growth as compared to shoot growth.

(iii) Root system architecture for efficient nutrient acquisition

Roots are highly plastic in response to environmental stimuli. Depending on the nutrient availability root architecture gets modified. Differential response has been observed for different nutrients or even different formulation of the same nutrient. Difference in root growth was observed when nitrogen was given as nitrate or ammonium. In general, local nitrate supply primarily affects lateral root elongation, whereas local ammonium supply primarily affects lateral root initiation. This can be explained in terms of variation in diffusion coefficient of the two different formulations. Diffusion coefficient of ammonium in soil is low as compared to nitrate which is a mobile soil resource. It has a higher diffusion rate in soil due to its negative charge.

Phosphorous is an immobile soil resource, hence in contrast to nitrate, the nutrient has low diffusion ability in the soil and it mostly exists in the top soils. Hence, growth of primary and lateral root growth is restricted in low phosphorus soils which may be due to reduced meristem activity. However, the root density was found to be higher and root hair growth was stimulated (Lambers *et al.,* 2006).

(iv) Root system architecture related to competition

Crop competition is also relevant for growth of the root system. When a plant's microenvironment is invaded by another due to plasticity the plant roots move away from the area of competition or their growth may be affected by severe resource crunch. This is especially true for high density planting. When planting density increases, light interception per plant decreases, resulting in a reduction in whole plant photosynthesis and biomass accumulation. Plasticity of the roots is such that it can adjust root length density and direction based on environmental cues. Therefore, the carbon allocated to the roots can be reduced, which can contribute to reduction in total length and density of roots under high density planting.

Modulation of plant architecture

Plant architecture can be modified by climatic factors, agronomic factors and by various human practices.

Climatic Factors

Light

Plants have evolved mechanisms that enable them to respond to growing environment. Leaf arrangement is critical for efficient capture of light and improving crop productivity. Canopy architecture is designed in such a way that plants are able to efficiently use resources depending on location. Photosynthetic pigments, chlorophylls and carotenoids, absorb light over most of the visible spectrum, although some green light is reflected or transmitted. Morphological modifications in petiole length, leaf size, flower diameter and leaf angle are found to vary with different spectral frequencies of light. Phytochrome plays a major role in light absorption and behaves differently under red (R) and far red (FR) light conditions. In dark, phytochrome is synthesized in the red light-absorbing (Pr) form (absorption maximum approx. 660 nm) and is generally regarded to be biologically inactive. A useful parameter to describe natural light environment is the ratio of photon irradiance in R to that in the FR (R: FR ratio). The R: FR ratio of daylight is around 1.15 and varies little with weather conditions or time of year. R: FR ratios underneath canopies of vegetation are typically in the range 0.05–0.7. In field crops, link between specific canopy architectural trait and plant productivity are poorly understood. When *Arabidopsis thaliana* was grown in low R:FR ratio, it characteristically displayed elongated stems/petioles, reduced leaf size, decreased chlorophyll content and early flowering often described as the R:FR ratio-mediated shade-avoidance response (Franklin and Whitelam, 2005).

Plant architecture is also reported to affect light absorption and photosynthesis in tomato. At high light intensities (summer) deeper penetration of light in the canopy improves crop photosynthesis, but not at low light intensities (winter). In particular, internode length and leaf shape affects vertical distribution of light in the canopy. A new plant ideotype of tomato with more spacious canopy architecture with long internodes and long and narrow leaves led to an increase in crop photosynthesis of up to 10 % (Sarlikioti *et al.*, 2011).

There is a quantitative relationship between yield and solar radiation at different growth stages in rice. Solar radiation at the reproductive stage has the greatest effect on grain yield; that at the ripening stage, the next highest effect; and that at the vegetative stage, an extremely small overall effect. Experiments conducted in genetically diverse rice varieties (Burgess *et al.*, 2017) showed that a plant type with steeper leaf angle allows more efficient penetration of light into lower canopy layers and this in turn contributes to greater photosynthetic potential (Burgess *et al.*, 2017).

Temperature

Influence of temperature on plant growth assumes significance in the current climate change scenario, because many phenological shifts in plant development have been linked to small increases in temperature. Temperature has marked effects on plant architecture by modulating timing of developmental events. It is seen that *Arabidopsis* grown at low temperatures showed dwarfed and compact rosette habit, with thicker leaves. With increasing temperature, a graded increase in stem elongation, leaf area and plant biomass is observed. In natural environments, it is likely that the difference between day and night temperatures (DIF) has an important influence on plant architecture. Greater elongation of stems and leaves is generally observed when daytime temperatures exceed night temperature (positive DIF), whereas the opposite effect results with a converse regime. Horticultural industry exploits alternating thermocycles (thermoperiodism) for regulating plant stature. In *Arabidopsis thaliana*, growth under high temperatures of 28°C result in striking elongation of stems and increased leaf elevation from the soil. In addition to displaying an elongated architecture, plants developed at high temperature produce fewer leaves than in 22°C grown controls. These leaves had reduced size and stomatal densities than those developed at 22°C.

Temperature regulates growth and development throughout the lifecycle of plants. In many species, a prolonged period of cold (stratification) is required to promote germination by stimulation of GA biosynthesis. Conversely, high temperature treatment of seeds can inhibit germination, through induction of Abscisic Acid (ABA). Timing of reproductive development is also temperature sensitive in many species. In winter annuals, a prolonged period of cold (vernalisation) is required to promote flowering in the following spring. In the model species, *Arabidopsis thaliana*, flowering time is accelerated by elevations in ambient temperature through induction of the floral integrator *FLOWERING TIME (FT)*.

Agronomic factors

Planting density

Various agronomic factors also play a major role in modification of plant architecture. Among these factors, plant density has a significant role. Currently, high density planting is gaining a lot of popularity. Major characteristics of high density planting is that the branches of the plants should not be interlocked so there should be maximum fruiting branches and minimum structural branches. Proper pruning should be done to overcome problems due to shade. Roots of the tree should spread freely without any competition Proportion between height of the plant and diameter should be maintained.

With lesser spacing there is an increase in plant height with increase in internodal length but there is a reduction in number of clusters and leaf area. This is due to reduction in amount of light received by plants. Even though there is high value of leaf area index fruit set rate is low because the optimum range of leaf area index is approximately equal to a value of 4.16 if it is higher or lower than that of this value it will affect the fruit setting process.

In a study conducted to examine the effect of plant density and cultivars on growth and yield of cowpea, it was observed that increasing plant population increased plant height and decreased number of leaves per plant and leaf area index (LAI). Increased plant density significantly increased seed yield per unit area, however the number of pods per plant, 100-seed weight, seed yield per plant and harvest index reduced with increased plant density (Naim and Jabereldar, 2010).

Manipulation by humans

Pruning

Pruning can be carried out in perennials to maintain an attractive plant shape. They bestow an attractive shape to ornamental plants, and in orchard trees it is adopted for initiation of new branches, maintaining tree architecture and also for activating budding and fruiting process This is accomplished through cutting and removing dead, broken, diseased and old branches. Pruning modifies light distribution within the canopy and enhances photosynthesis, increases stem potential, induces canopy transpiration and improves water status. Woody plants are pruned to desired size and shape by judicious removal of plant parts so as to promote efficiency and performance and also to retain a certain structural morphology. Pruning in plant parts is based on the following broader principles.

1. Overcoming apical dominance
2. Changing phase of growth
3. Maintaining a balance between root and shoot growth
4. Relating to environmental condition

Ornamental plants are pruned to improve their aesthetic quality, but fruit trees are pruned to improve fruit quality by encouraging an appropriate balance between vegetative (wood) and reproductive (fruiting) growth. Annual pruning of fruit trees reduces yield, but enhances fruit quality. Pruning increases fruit size because excess flower buds are removed and pruning encourages growth of new shoots with high-quality flower buds. It also improves light penetration into canopy, growth, flower-bud development, fruit set, and red color development. Pruning also makes the canopy more open and improves pest control by allowing better spray penetration; air movement throughout the canopy, reduces humidity and improves drying conditions (Suxia *et al.*, 2009). It is considered a very important cultural operation in commercial cultivation of many tree crops. In nutmeg, pruning helps in the development of a suitable canopy structure which increases fruit bearing capacity through synchronized flowering. This in turn facilitates easy harvest operations and improves the benefit cost ratio (B/C ratio) of the produce (Aravind *et al.*, 2019).

Effect of pruning on growth of bougainvillea

Style of pruning and the age of the crop helps in the circulation and gathering of nutrient elements which influences regrowth. Saifuddin *et al.*, 2010 conducted a number of pruning studies with the genus *Bougainvillea.* There were four treatments. Control plant (T_1) was not pruned within six months. The other treatments included partial pruning in all branches except one branch (T_2), complete pruning by cutting all primary and secondary branches (T_3) and frequent pruning of all branches (T_4), 4 cm from shoot apex, at 30 days interval throughout the experimental period.

Observations focused on initiation of new flowering shoots. The shoot growth was found to be lower in the non-pruned plants, which may be due to short-term reduction of cell activity of older shoots, causing a decline in flower production. In addition, due to extended vegetative phase, available nutrient reserves may be expended for vegetative growth rather than for floral initiation. It was seen that pruning helped flower initiation by creating or increasing the availability of metabolic sinks. Flower recovery was dependent on pruning, especially pruning-position, pruning-height and the timing of plant phase. These results were substantiated by the effect of pruning on the physiological and biochemical components of the plants such as chlorophyll a and b contents, quantum yield, sugar content, biomass and stomatal conductance on flowering process.

Frequent pruning of *B. glabra* plants gave the highest quantum yield, chlorophyll a and b contents, and maximum flower initiation per plant compared to that of non-pruned plants (control). Sugar content in pruned plants decreased, probably due to their utilization for development of new flowering shoots and their growth. Minimum branch fresh weight and low potassium content was observed in the completely pruned plants (T_3). Observations also indicated that different pruning treatments influences the physiological and biochemical activity of plants and modifies root and shoot initiation pattern. In case of completely pruned (T_3) plants, the shoot growth rate was minimum due to low availability of potassium, sugar, nutrients and lower root growth. Tertiary branch initiation was found to be more predominant in the non-pruned condition. The above findings have given a better understanding of the effects of different pruning on the growth and development of *B. glabra.* Hence, it is suggested that frequent pruning can be used to trigger flower shoot initiation and maintain plant growth.

Moreover, pruning increases the supply of cytokinins from the roots as seen by an increase in concentration of the hormone in the remaining above-ground tissue. The increase of hormone levels is probably responsible for stimulating cell division, new shoot formation and ultimately more flower per branch and frequent flower bud initiation. Frequent pruning took lesser time to induce flowers by continuously stimulating cytokinin production for bud formation and therefore flower number per branch, bract length and weight remained unchanged from the first to final season. However, in the case of partial and complete pruning, number of flowers per branch was low in first and final seasons due to the prolonged vegetative stage for shoot initiation.

Pruning and canopy architecture in cashew

High density planting of fruit orchards is a commercial activity which is gaining a lot of attention in current day agricultural practices. Maintaining an ideal architecture of plants by rigorous pruning is a major factor that decides the success of any plantation. A study undertaken by Onguso *et al.,* 2004 showed that under high density orchard system in a cashew plantation, annual pruning of exhausted growth helps the plants to remain productive. Annual growth and yield per unit area of pruning responsive cashew varieties was superior over plants planted in higher spacing with lesser pruning.

Hormonal influence on pruning

Auxins produced in terminal buds suppresses growth of side buds and stimulates root growth. Cytokinins produced in growing tips of root stimulate shoot growth. Pruning a newly planted tree removes auxin, slowing root regeneration. Heading cuts (removal of a branch tip) releases apical dominance caused by auxins from terminal buds. This allows side shoots to develop and branches become bushier. On the other hand, thinning cuts remove a branch back to the branch union (crotch). This type of cut opens the plant to more light. Most pruning should be limited to thinning cuts (Onguso *et al.*, 2004)

Bending in plants

Bending is a super cropping procedure similar to low stress training. Normally plants bend towards sun since sunlight provides the assimilatory power for growth. Pruning and bending can be applied to various practices like bonsai, pleaching, pooktre, topiary etc. (Coutand *et al.*, 2000). Bending the plant in a horizontal position also allows the plant to grow buds that will grow upward. When being bent, the plant is getting more sunlight than it was since the obstruction of leaves of the upper branches are removed, so the sides of the plant that were not having direct exposure will now absorb direct sunlight. Another benefit acquired from bending is that growth suppressing hormones are deactivated, and the ability to manipulate plant size shape is given to the grower. Bending is a lot less stressful and riskier than pruning. If necessary, the bent branches can be supported with a piece of thread or any similar lining material so the plant would not revert to its original shape.

a. Bonsai

Bonsai is the art of developing a trees or plants into an aesthetically appealing shape after growing plants in suitable containers, and by pruning and training them. The word bonsai is made of two contributory words "Bons" meaning shallow pot and secondarily "Sai" meaning planting which can be translated as "tray planting". There are many myths which are associated with bonsai. Bonsais are not genetically dwarfed plants but they are kept small by cruelty. Bonsai can be developed from seeds or cuttings, from young trees or from naturally occurring stunted trees transplanted into containers (Fig. 1). The most common bonsai range in height from 5 cm to 1 m. To keep them small they have to be trained by pruning roots, periodic repotting, pinching off new growth and making wiring of the branches and trunk which gives desired shape to the trunk and branches of plants. Hybrid dwarfs do not fall into this category; only ordinary plants kept dwarf can be called as "Bonsai". Small leaved varieties are most suitable, but essentially any plant can be used, regardless of its size in the wild. In Japan, varieties of Japanese maple, pine, Azalea, Camellia, bamboo and plum are most often used. Two basic things to maintain the dwarf nature of plants are wiring and pruning (Kumar and Dwivedi, 2011).

Wiring

Wiring is a crucial technique to train and style Bonsai trees. Copper wire is used very commonly for training and bending of bonsai trees. The gauge of the wire will depend on the specimen of the plants. During wiring, precaution is taken to see that first the wire is bound round the trunk several times and then around the branches. The wire coil should be spaced evenly by about 0.5 to 0.06 cm. Rewinding is to be done every six months.

Fig. 1: Bonsai collection from Joseph P J: **(a)** *Adenium obesum*; **(b)** *Ficus retusa*; **(c)** *Triphasia trifolia.*

Pruning and Pinching

Bonsai tree requires at least 3 to 6 h of adequate quantity of sun light daily to grow properly. The last exercise is trimming the particular crown of the bonsai. Main attraction of bonsai trees are their physical and morphological appearance. To sustain its gorgeous look regular trimming is essential. In case of saplings regular pruning is important to minimize expansion. Other methods of pruning such as shoot pruning, leaf pinching and root pruning also helps to retain the dwarf stature of the plants.

Physiology of bonsai

Shoot and Root Balance

It is important to maintain a proper root shoot balance for these miniature plants to retain the morphology of the species. Generally, shoots and roots grow in opposite direction by maintaining symmetry away from the center. Since cell division is restricted to the meristematic regions, cell division is slow as the existing cells get larger and stretch out on a vertical axis. Growth and maturation take place in areas where shoot and root are closest. The roots and shoots will maintain a steady, proportional relationship. To retain the miniature stature of the plant, it is important to balance the water and nutrients supplied to them. To provide extra support for the growing tree, the roots become stout and large. Moreover, in a stunted state, the root acts as a secondary sink and there can be more translocation of food and nutrients to the roots which activates root growth. Hence, root pruning is an important procedure, wherein nearly one third of the roots have to be pruned and removed each year. Fig. 2a shows the ideal method of root removal. The second method shown in Fig. 2b is more cumbersome and chances of root damage are also more in this case.

Choosing the container is also important because if the base is overly small root growth will be inhibited and the shoot will quit growing as the root system will not be able to support a larger shoot. This can also happen if the roots are damaged during repotting. This can be detected when the plants show signs of compensation such as losing leaves or wilting. Once the roots are given adequate space and time for development the shoots can regain growth.

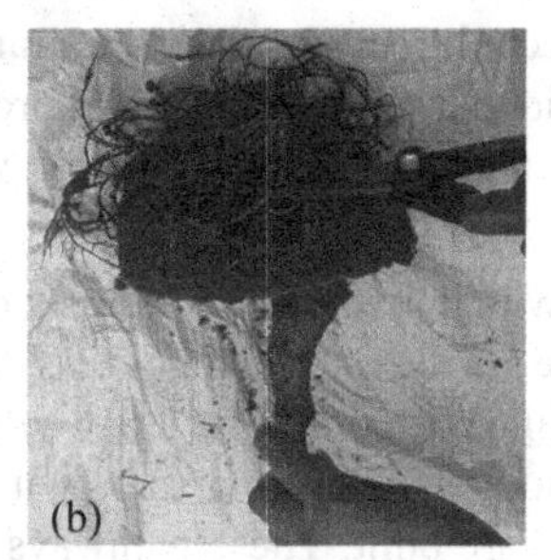

Fig. 2: Root pruning in Bonsai production (*Image courtesy:* Joseph P J)

Secondary growth

Secondary growth is not a common feature in all plants but for bonsai production plants which have secondary thickening are normally chosen. Secondary growth is the thickening of the stem, as stem growth increases there is higher demand for water and nutrients this requires larger channels for transport which leads to increase in the diameter of the stem. To keep up with that increased demand, the roots also has to grow. Secondary growth occurs in the cambium it increases thickness of the cell walls and eventually dies. This tissue hardens and form supportive tissue for the stem. Secondary growth is seen both in roots and shoots It helps in developing a thickened trunk and an overall tapered form.

Hormonal balance

Hormones like auxin and cytokinin play a major role in maintaining the dwarf nature of bonsai plants. Auxin which is synthesised in shoot tip will move through phloem and reach the roots where it induces production of a greater number of lateral roots. Cytokinin which is synthesized in roots will move through xylem and reaches the shoot tip where it increases production of a greater number of branches. Increase in thickness is due to secondary growth.

Role of growth retardants in canopy management of Bonsai

Growth retardants are those chemical substances that retard stem elongation without causing any malformation of plants eg. Phosfon-D (2, 4-dichlorobenzyltributyl phosphonium chloride), cycocel or CCC (2-chloroethyl trimethylammonium chloride) and B-nine (N-dimethyl amino succinamic acid). These chemicals have been reported to be effective on a large number of plants. Cycocel (liquid) is used as foliar spray or soil drench. Phosfon-D and cycoel dust can be applied in soil, while B–nine is used as foliar spray. In case of perennial plants, chemicals are used when new shoots on pruned plants attain 5-10 cm length.

b. Pleaching

Pleaching is a style of growing trees in a line, usually straight, with the branches of the tree tied together and clipped to form a flat plane above the bare trunk. The branches are tied onto canes or wire to make tiers, and are then regularly pruned to keep their shape. Sometimes they naturally graft themselves onto one another. Trees whose branches need to be trained are tied to a bamboo frame. It is essential to imagine what they will be like in five, 10 and 20 years' time. Hence it is necessary to leave lots of space for them to grow. Selected trees should have straight trunk with same girth and regular pruning should be done in order to maintain proper shape. This technique is adopted as live boundaries instead of building walls for partition.

c. Pooktre

It is a gradual shaping method, which involves the shaping of trees as they grow along predetermined designs. Tree lore are the rules that governs the responses of a particular species of a tree to outside stimuli. With this knowledge a design is created that works with the trees' architecture. Gradual tree shaping starts with a supporting framework, into which the growth pathways are built. These pathways are either on wooden jigs or a shaped wire pathway. Seedlings or small cuttings 7-30 cm (3-12 in) tall are then planted into this framework. The actual shaping of new growth happens with training of the shaping zone. This training is a daily or weekly shaping of young growth. The time required to grow a tree along the length of the design depends on the size of the project. Once the trees have grown to their full length, they are allowed to thicken with time. Techniques like pruning, grafting, and training the shaping zone, are practiced to achieve a particular shape. Main drawback of this technique is that it can take from one growing season to several years to develop the required shape and thickness of the design. A few representative images of Pooktre designs and a typical shaping zone depiction are given in Fig. 3.

Fig. 3: Representative Pooktre designs using *Prunus myrobalan.*
Image courtesy and details: Peter Cook and Becky Northey-the founders of Pooktre.

(a) Garden chair: The shape was grown within first two years, but it took 5 years to be thick enough to be used as a real chair. The tree was 18 years old at the time of this photo. **(b) Person tree:** This design was inspired by Japanese sumo wrestlers. The shape was grown in the first year. This photo is of a 12-year-old tree. **(c) Shaping Zone** is different for every species of trees. This image is of *Prunus myrobalan's* shaping zone. In the shaping zone, cells do not have a fixed orientation and so can be flexed or carefully removed without damage.

d. Topiary

Topiary is the horticultural practice of training live perennial plants by clipping the foliage and twigs of trees, shrubs and subshrubs to develop and maintain clearly defined shapes, whether geometric or fanciful. The term also refers to plants which have been shaped in this way. Plants used in topiary are evergreen, mostly woody, have small leaves or needles, produce dense foliage, and have compact and/or columnar (e.g., fastigiate) growth habits. Common species chosen for topiary include cultivars of European box (*Buxus sempervirens*), arborvitae (*Thuja* species), bay laurel (*Laurus nobilis*), holly (*Ilex* species), myrtle (*Eugenia* or *Myrtus* species), yew (*Taxus* species), and privet (*Ligustrum* species), *Phyllanthus sp. etc.* (Fig. 4). Shaped wire cages are sometimes employed in modern topiary to guide untutored shears, but traditional topiary depends on patience and a steady hand; small-leaved ivy can be used to cover a cage and give the look of topiary in a few months. H is a simple form of topiary used to create boundaries, walls or screens.

Fig. 4: Topiary of *Phyllanthus myrtifolius* (Wight) Mull. Arg. (Myrtle-leaved Leaf-flower) Image courtesy: Anil A S

Molding of fruits

Molding means giving particular shape to fruits and vegetables by using different plastic or woody molds. Fruits and vegetables like tomatoes, cucumbers, lemons and mandarin oranges are commonly used for this technique. It is done by placing small immature fruits and vegetables in molds shaped according to our preference. These fruits grow slowly and when they attain the required shape they can be harvested.

Physiology of fruit architecture

Ovary wall fruit comprises of outer exocarp and mesocarp which is separated by vascular bundles. Inner pericarp tissue comprises of endocarp with placenta and ovule. Size and shape of fruits is determined mainly by number of cells and direction of cell division. Cell division at vascular bundles and placenta will change the shape. Some of the genes also plays a major role in maintaining the shape of fruits like *SUN and OVATE* genes are mainly for fruit elongation whereas *LOCULE NUMBER (LC) and FASCIATED (FAS)* for maintaining locule number and flat fruit shape (Monforte *et al.*, 2013).

Advantages of modification of plant architecture

- Accommodates more plants/unit area
- Increases light harvesting capacity
- Increases flower initiation and fruit set
- Improves productivity
- Improves appeal of fruits and vegetables for market preference.
- Facility for packing and transportation
- Improves aesthetic quality of plants

Conclusion

Architecture of a plant depends on the nature and relative arrangement of each of its parts, which is the expression of equilibrium between endogenous growth processes and exogenous constraints exerted by the environment. The aim of architectural analysis is to identify and understand these endogenous processes and to separate them from plasticity of their expression resulting from external influences.

Architectural analysis is essentially a detailed, multilevel, comprehensive and dynamic approach to plant development. Despite their recent origin, architectural concepts and analysis methods provide a powerful tool for studying plant form and ontogeny, by precise morphological observations and appropriate quantitative methods of analysis. Recent researches in this field have greatly increased our understanding of plant structure and also helped to identify several morphological criteria for development and this has led to the establishment of a real conceptual and methodological framework for plant form and structure analysis.

Acknowledgement

The authors thankfully acknowledge Peter Cook and Becky Northey, the founders of Pooktre, for sharing their designs and the associated technical details, and Joseph P J for sharing the technical details of his Bonsai collection.

References

Amaya, I., Ratcliffe, O. J., and Bradley, D. J. 1999. Expression of the CENTRORA DIALIS (CEN) and CEN-like genes in tobacco reveals a conserved mechanism controlling phase change in diverse species. *Plant Cell* 11: 1405-1417.

Aravind, S., Kandiannan, K., Rema, J., S. Ankegowda S. J., and Senthil Kumar, R. 2019. Enhancement of yield in nutmeg (*Myristica fragrans* Houtt.) through pruning. *Journal of Plantation Crops* 47(2): 121-123.

Barthelemy, D. and Caraglio, Y. 2007. Plant architecture: a dynamic, multilevel and comprehensive approach to plant form, structure and ontogeny. *Ann. Bot.* 99(3): 375-407.

Bayer, E. M., Smith, R. S., Mandel, T., Nakayama, N., Sauer, M., Prusinkiewicz, P., and Kuhlemeier, C. 2009. Integration of transport-based models for phyllotaxis and midvein formation. *Genes Dev.* 23: 373-384.

Benfey, P. N. and Scheres, B. 2000. Root development. *Curr. Biol.* 10: R813–R815.

Benjamins, R., Quint, A., Weijers, D., Hooykaas, P., and Offringa, R. 2001. The PINOID protein kinase regulates organ development in Arabidopsis by enhancing polar auxin transport. *Development* 128: 4057–4067.

Benková, E., Michniewicz, M., Sauer, M., Teichmann, T., Seifertová, D., Jürgens, G., and Friml, J. 2003. Local, efflux-dependent auxin gradients as a common module for plant organ formation. *Cell* 115: 591-602.

Berling, J. D. and Fleming, J. A. 2007. Zimmermann's telome theory of megaphyll leaf evolution: a molecular and cellular critique. *Current Opinion in Plant Biology* 10(1):4-12.

Busov, V. B. 2018. Manipulation of growth and architectural characteristics in these for increase woody biomass production. *Front. Plant Sci.*

Burgess, A. J., Retkute, R., Herman, T., and Murchie, E. H. 2017. Exploring relationships between canopy architecture, light distribution and photosynthesis in contrasting rice genotypes using 3D canopy reconstruction. *Front. Plant Sci.* https://doi.org/10.3389/fpls.2017.00734.

Cao, H. and Chen, S. 1995. Brassinosteroid-induced rice lamina joint inclination and its relation to indole-3-acetic acid and ethylene. *Plant Growth Regulation*16: 189–196.

Coen, E. S. and Romero, J. M. 1994. Evolution of flowers and inflorescences. *Development.* 13(3): 107-116

Coutand, C., Julien, J. L., Moulia, B., and Guitard, D. 2000. Biomechanical study of the effect of a controlled bending on tomato stem longation. *J. Exp. Bot.* 51: 1813-1824.

Dardick, C., Callahan, A., Horn, R., Ruiz, K. B., Zhebentyayeva, T., Hollender, C., Whitaker, M., Abbott, A., and Scorza, R. 2013. *PpeTAC1* promotes the horizontal growth of branches in peach trees and is a member of a functionally conserved gene family found in diverse plants species. *Plant J.* 75(4): 618-30.

Davies, C. R., Seth, S. K., and. Wareing, P. F. 1966. Auxin and Kinetin Interaction in Apical Dominance. *Science* 151*: 468-469*

Douady, S. and Couder, Y. 1996. Phyllotaxis as a dynamical self organizing process part II: The spontaneous formation of a periodicity and the coexistence of spiral andwhorled patterns. *Journal of Theoretical Biology* 178(3):275–294.

Franklin, A. K. and Whitelam, C. G. 2005. Phytochromes and shade-avoidance responses in plants. *Ann. Bot.* 96: 169-175.

Halle. F and Oldeman R. A. 1970. Dynamique de croissance des arbres tropicaux. *Ann. Bot.* 14(2): 169-175.

Heisler, M. G., Ohno, C., Das, P., Sieber, P., Reddy, G. V., Long, J. A., and Meyerowitz, E. M. 2005. Patterns of auxin transport and gene expression during primordium development revealed by live imaging of the Arabidopsis inflorescence meristem. *Curr. Biol.* 15: 1899-1911.

Hofmeister 1868. Allgemeine morphologie der gewachse. In A. de Bary, T. H. Irmisch, & J. Sachs (Eds.), Handbuch der Physiologischen Botanik pp. 405–664. Leipzig:

Hollender, C. A. and Dardick, C. 2015. Molecular basis of angiosperm tree architecture. *New Phytologist* 206, 541–556

Huang, S., Gao, Y., Li, Y., Xu, L., Tao, H., and Wang, P. 2017. Influence of plant architecture on maize physiology and yield in the Heilonggang River valley. *Crop J. 5*(1): 52-62.

Kang, J., Park, J., Choi, H., Burla, B., Kretzschmar, T., Lee, Y., and Martinoia, E. 2011. Plant ABC Transporters. *The Arabidopsis Book 9*:e0153.

Křeček, P., Skůpa, P., Libus, J., Naramoto, S., Tejos, R., Friml, J., and Zažímalová, E. 2009. The PIN-FORMED (PIN) protein family of auxin transporters. *Genome Biol.* 10:249.

Kumar, P. and Dwivedi, P. 2011. Bonsai: symbol of culture, ideals, money and beauty. *Int. Jr. Agril. Env. Biotech.* 4(2) : 115-118.

Lambers, H., Shane, M. W., Cramer, M. D., Pearse, S. J., and Veneklaas, E. L. 2006. Root structure and functioning for efficient acquisition of phosphorus: Matching morphological and physiological traits. *Annals of Botany* 98(4): 693–713.

Lynch, J. 1995. Root architecture and plant productivity. *Plant physiol.* 109(1): 7.

Monforte, A. J., Diaz, A. I., Delgado, A. C. and Knapp, E. D. 2013. The genetic basis of fruit morphology in horticultural crops: lessons from tomato and melon-a review. *J. Exp. Bot.* 10: 1-13.

Naim, A. M. and Jabereldar, A. A. 2010. Effect of plant density and cultivar on growth and yield of cowpea (Vigna unguiculata L. Walp). *Aust. J. Basic and Appl. Sci.* 4(8): 3148-3153.

Onguso, J. M., Mizutani, F. and Hossain 2004. Effects of partial ringing and heating of trunk on shoot growth and fruit quality of peach trees. *Bot. Bull. Acad. Sin.* 45: 301-306.

Pineiro, M. and Coupland, G. 1998. The control of flowering time and floral identity in Arabidopsis. *Plant Physiol.* 117(1): 1-8.

Reinhardt, D. and Kuhlemier, C. 2002. Plant architecture. *Eur. Mol. Biol. Rep.* 3(9): 846-851.

Reinhardt, D., Pesce, E.-R., Stieger, P., Mandel, T., Baltensperger, K., Bennett, M., Traas, J., Friml, J., and Kuhlemeier, C. 2003. Regulation of phyllotaxis by polar auxin transport. *Nature* 426: 255-260.

Saifuddin, M. Hossain, A. B. M. S., Osman, N., Sattar, M. A., Moneruzzaman, K. M. and Jahirul, M. I. 2010. Pruning impacts on shoot-root-growth, biochemical and physiological changes of Bougainvillea glabra. *Aus. J. Crop Sci.* 4(7): 530-537.

Sarlikioti, V., de Visser, P. H. B., Buck-Sorlin, G. H., and Marcelis, L. F. M. 2011. How plant architecture affects light absorption and photosynthesis in tomato: towards an ideotype for plant architecture using a functional–structural plant model. *Annals of Botany* 108(6): 1065–1073.

Schoute, J. C. 1913. Beitrage zur Blattstellunglehre. I. Die Theorie. Recueilde Travaux *Botaniques Neerlandais* 10: 153–339.

Snow, M. and Snow, R. 1962. A theory of the regulation of phyllotaxis based on *Lupinus albus*. *Philosophical Transactions of the Royal Society B: Biological Sciences* 244(717): 483–513.

Stecconi, M. 2006. Variabilidad arquitectural de especies nativas de Nothofagus de la Patagonia (*N. antarctica, N. pumilio, N. dombeyi*) Argentina: Regional Universitario Bariloche – Universidad Nacional del Comahue; PhD thesis in Biology, Centro.

Steffens, B. and Rasmussen, A. 2016. The Physiology of Adventitious Roots. *Plant Physiology* 170(2): 603-617.

Sussex, I. M. and Kerk, N. M. 2001. The evolution of plant architecture. *Curr. Opinion Plant Biol.* 4(1): 33-37.

Suxia, X., Qingyun, H., Qingyan, S., Chun, C., Brady, A. V. 2009. Reproductive organography of *Bougainvillea spectabilis*. *Willd. Sci. Hortic.* 120: 399-405.

Swarup, R. and Péret, B. 2012. AUX/LAX family of auxin influx carriers-an overview. *Frontiers in Plant Science*, 3: 225.

Yu, B., Lin, Z., Li, H., et al., 2007. TAC1, a major quantitative trait locus controlling tiller angle in rice. *The Plant Journal* 52: 891–898.

Zhang, L. Y., Bai, M. Y., Wu, J., et al., 2009. Antagonistic HLH/bHLH transcription factors mediate brassinosteroid regulation of cell elongation and plant development in rice and Arabidopsis. *The Plant Cell* 21: 3767–3780.

Zhao, S. Q., Hu, J., Guo, L. B., Qian, Q., and Xue, H. W. 2010. Rice leaf inclination2, a VIN3-like protein, regulates leaf angle through modulating cell division of the collar. *Cell Research 20*: 935–947.

7

Plant Neurobiology A Paradigm Shift in Plant Science

Sreepriya S* and *Girija T

Department of Plant Physiology, College of Horticulture, Vellanikkara Kerala Agricultural University, Thrissur

Introduction

Plants are considered as immobile organisms with poor sensitivity and limited ability to respond. Within the plant kingdom, *Mimosa pudica* (also called the sensitive plant), *Drosera* (*sundews*), *Dionea muscipula* (flytraps) and tendrils of climbing plants which showed rapid and purposeful movements were considered as exceptions. These sensitive plants attracted attention of outstanding pioneer researchers like Pfeffer (1845-1920), Burdon-Sanderson (1828-1905), Darwin (1809-1882), Haberlandt (1854-1945) and Bose (1858-1937). Stahlberg, 2006 reported the presence of various mechanoreceptors in such species which could trigger action potentials (APs) which activated these sudden movements.

The first extracellular recording of a plant action potential (AP) development was initiated by Charles Darwin; later, the animal physiologist Burdon-Sanderson in 1873 performed similar studies on leaves of the Venus flytrap (*Dionea muscipula* Ellis). Haberlandt (1884) demonstrated that phloem strands were the actual paths for excitation by destroying the external, non-woody part of vascular bundles. This notion was confirmed in *Mimosa* and other plant species by a number of recent studies. Houwink (1935) carried out experiments with two species of *Vitis* (grape) and obtained electrical fluctuations which was similar and comparable with the action potentials he obtained with *Mimosa*. In 1907, Bose, used a D'arsonval galvanometer to demonstrate propagating electrical effects in different genera of plants such as *Ficus*, *Artocarpus*, *Cucurbita*, *Corchorus*, "fern" *etc.* APs have their largest amplitude in phloem sieve cells and cells adjacent to this area (Sibaoka, 1969). Other studies found that AP-like signals are propagated with equal rate and amplitude throughout the cells of vascular bundles. Discovery that normal annual plants such as pumpkins had propagating action potentials (Opritov and Pyatygin, 1989) just as the esoteric "sensitive" plants was a scientific breakthrough with important consequences. However, work on plant neurobiology was accepted only as a pseudoscience for a long time by the scientific community. In 2003, scientists and educators such as Stefano Mancuzo and Elizabeth Van Volkenburg, from the University of Washington and Frantisek Baluska from University of Bonn together established the 'Society for Plant Neurobiology' which deals with signaling and behaviour of plants.

In general, communication in plants can be categorized into

1. 'within plant',
2. 'between plants'
3. Communication with beneficial organisms

Case studies on plant communication

Communication within plant

Environmental factors were found to influence the development of stomatal pores in leaves. Stomata are microscopic pores on the surfaces of leaves, the number and density vary in response to changes in environmental conditions, such as carbon dioxide concentration and light. Lake *et al.,* (2001) showed that mature leaves of *Arabidopsis thaliana* detect and transmit information regarding the external environmental factors to new leaves of the same plant, producing an appropriate adjustment of stomatal development. This was demonstrated through an experiment conducted on *Arabidopsis* (Columbia, Col-0) plants which were grown up to five leaf stage under ambient CO_2 condition which was 360 ppm. One set of the plants were than enclosed in transparent airtight cuvettes under CO_2 concentrations of 720 ppm. These were maintained for 7 to 9 days until the next five leaf insertions (5th leaf to 13th leaf) had matured, the last three of which were investigated for stomatal density (no. of stomata per mm^2) and index ((no. of stomata/no. of stomata & no. of epidermal cells) $\times$ 100). In the cuvette, where the mature leaves were exposed to high concentration of CO_2 (720 ppm), the newly expanding leaves had reduced stomatal index and density, when compared with control plants grown entirely at ambient CO_2. In a reverse experiment, where mature leaves alone were kept under ambient CO_2 (360 ppm) and the plant was exposed to 720 ppm CO_2, stomatal density and index increased in new leaves in response to decreased CO_2 around the mature leaves.

In these experiments, both abaxial (lower) and adaxial (upper) leaf surfaces responded in a similar manner, indicating that CO_2 concentration was detected by mature leaves which signalled to expanding leaves to induce an appropriate developmental response. This was one of the earliest demonstrations that proved the role of mature leaves in detecting CO_2 concentration and also its ability to transmit a long-distance signal that controls stomatal development in young leaves. Expanding leaves appeared to have no capacity to detect ambient CO_2 concentration or to respond to it directly by altering stomatal initiation.

Communication between plants

In nature, several stress factors such as disease, injury, herbivory, extreme heat/cold *etc.* hinder plant growth and development. So as a survival strategy plants will have to adjust their physiological state either in response to, or in preparation for, such threats (Vickers *et al.,* 2009). It is now known that plants are capable of communicating such information with the help of volatile organic compounds (VOCs).

These compounds are produced by plants under different circumstances. In the case of flowers, VOCs are produced to attract pollinators and ensure self pollination; contrary to our concept, VOCs produced by the plants are more than just a scent

(Koptur, 1992). In a damaged plant, VOCs are also used as non-volatile signals to transmit messages within the plant itself. Airborne signals are diffused to reach neighbouring undamaged plants giving them a chance to strengthen their own defense system. The receivers are not limited to con species alone. Natural enemies can also catch the signals and locate the place of battle (Hare, 2011).

VOC mediated defense against herbivores

Poplar and sugar maple trees were seen to accumulate phenolics and tannins when situated close to damaged trees. This observation paved the way for further research in this area. Methyl jasmonate (MeJA) emitted by sagebrush (*Artemisia tridentata*) due to herbivore attack was the first compound detected to impart resistance in intact plants, by increasing the proteinase inhibitor production (Farmer and Ryan, 1990). Later a number of other VOCs emitted by damaged plants were found to influence the receiver plants.

In a study conducted by Hirokazu *et al.,* (2012), to examine the effects of wound-induced VOCs on the biosynthesis of pyrethrins, it was seen that the amount of pyrethrin increased in intact young seedlings of Pyrethrum daisy (*Tanacetum cinerariifolium*; earlier species name: *Chrysanthemum cinerariaefolium*) by placing them in the vicinity of wounded seedlings. These insecticidal metabolites imparted tolerance to the plant species. Elucidation of VOCs emitted by wounded seedlings of *T. cinerariifolium* by GS-MS showed the presence of (Z)-3-hexenal, (E)-2-hexenal, (Z)-3-hexen-1-ol, (Z)-3-hexen-1-yl acetate and (E)-β-farnesene. It was also seen that the blend ratio of various organic volatiles varied dynamically with time after wounding. It was possible to artificially activate the biosynthesis of pyrethrin when these five components were mixed together in a ratio similar to that observed 35–60 min after wounding. However, a 10-fold increase and a decrease to 1/10 the concentration of VOCs caused a marked reduction in gene expression of 1-deoxy-D-xylulose 5-phosphate synthase (DXS), chrysanthemyl diphosphate synthase (CPPase) and allene oxide synthase (AOS), which were involved in the biosynthesis of pyrethrin. Moreover, it was also observed that only when all the 5 different components were mixed together was pyrethrin synthesis effectively achieved. Eliminating just one component from the five-VOC mixture resulted in reduced gene expression of 13-lipoxygenase as well as DXS, CPP and AOS, demonstrating that both the concentration and blend ratio play an important role in establishing plant-plant communications.

These studies revealed that individual VOCs, their blend and concentration are important in plant to plant communication. It was also noted that individual VOCs may not be specific for the plant species but their blend ratios imparts specificity to the VOCs in plant-plant communications between con species. When plants rely only on specific VOCs they may not be able to respond to herbivores that target a broad range of plant species. Plants also seem to eavesdrop on the herbivore-induced VOCs from other species.

Plant-Parasite communication

Parasitic plants are important components of both natural and agricultural ecosystems and have considerable influence on structure and dynamics of the communities they inhabit (Press and Phoenix, 2005). Flowering plants in the genus *Cuscuta* are obligate parasites with little photosynthetic capability; they obtain nutrients by attaching to aboveground shoots of other plants. Post germination, *C. pentagona* seedlings establishes contact with the host by following a rotational growth pattern (circumnutation). In addition to the earlier knowledge on the role of host secondary metabolites to influence below ground growth of parasitic plants that attach to host roots, it was also reported that host-derived chemicals also induce haustorial development of these parasites (Yoder, 1997). In a classic experiment conducted by Runyon *et al.,* (2006), on host identification and interaction of *C. pentagona,* they were able to prove that a large proportion of parasite seedlings grew more or less directly toward the target host plant. In order to confirm the involvement of host-plant volatiles as possible cues responsible for eliciting such a response from the parasite, they substituted the host with volatiles extracted from 20 day old tomato plants. Growth response to extracted volatiles was similar to that observed in response to whole plants, thus confirming the role of volatiles in host detection by the parasitic plant *C. pentagona.*

Case studies on plant behavior

One of the characteristic features of any living organism is its behaviour. Plants can accurately calculate their circumstances, use intricate cost-benefit analysis, and take defined actions to mitigate and control diverse environmental emergencies. They are capable of a refined self- and non-self-recognition, exhibit territorial behaviours and exhibit complex communication skills. Communication and signaling in plants encompass both chemical and physical pathways. Plants interact with animals wherein they attract them with colourful flowers or fleshy fruits to make sure their flowers get pollinated and their seeds dispersed. They provide sugary nectars to recompense them for their protective services. Plants exhibit different cooperative or antagonistic behaviours according to the degree of relativeness among them.

Trans-generation memory of stress in plants

Plants are influenced by abiotic and biotic environmental factors wherein, changes in plant physiology and genome dynamics are exhibited as resistance responses. These include the activation of transposable elements by abiotic and biotic stress conditions, induction of mutations by chemical and physical agents, and enhancement of homologous recombination by elevated temperatures or ultraviolet-B (UV-B) (Ries *et al.,* 2000). The genomic flexibility shown by plant genomes evidenced by changes at the sequence level of the affected genes, in response to pathogen attack is very exciting (Lucht, 2002). The influence of these changes have in evolutionary terms, however, remained poorly understood, because most changes were detected in somatic tissue and not considered in further generations. In plants, the reproductive cell-lineage emerges from somatic tissue late in development, thus some genomic changes acquired during the life of a plant can be transmitted to the next generation. The frequency of occurrence of genetically fixed mutation was reproducibly elevated in the progeny of UV-B or pathogen-treated plants.

Molinier *et al.,* (2006) measured the rate of homologous recombination in the untreated offspring of plants exposed to different ultraviolet-C (UV-C) exposure conditions, using *A. thaliana* plants harbouring β-glucuronidase (GUS)-based constructs, in which truncated but overlapping parts of the gene was used to quantify somatic homologous recombination. Using this assay, the influence of UV-C was tested in *A. thaliana* plants that carried the recombination reporter in different relative orientations of the GUS sequence fragments. Treatment with UV-C stimulated the level of homologous recombination in the transformed plants. The homologous recombination frequency was measured in four subsequent generations S_1-S_4. Somatic homologous recombination was persistently found to be increased up to the selfed S_4 progeny of plants that were only treated with UV-C in S_0. The epigenetic changes leading to enhanced homologous recombination frequencies is thus stable for at least four generations. This may be the underlying molecular mechanism that causes the observed trans-generational 'memory'.

Problem solving ability of plants

Plants have to control consumption of food reserves produced by photosynthesis to prevent starvation during periods when food acquisition is not possible. Dark/light cycles of plants and ability to survive extreme climatic conditions like winter and snowfall come under this category. A classic study was conducted by Yazdanbakhsh *et al.,* in 2011; they studied the response of light and dark cycles in mutants of *Arabidopsis thaliana* plants. Photosynthesis utilizes solar energy for carbon assimilation during day and at night when light energy is not available, they utilize stored carbohydrates mostly starch to continue metabolism and growth. It was seen that *A. thaliana* mutants which had defects in either accumulation or degradation of starch showed symptoms of starvation.

It was observed that more than half of the starch produced during photosynthesis is stored in the form of crystalline starch granules in the chloroplast of *A. thaliana* and leaf starch content showed a linear increase during day time. During night there was a linear decrease in starch content with time and by dawn 95% of available starch was already utilized by the plants (Gibon *et al.,* 2004). This was even true when darkness came unexpectedly early (Graf *et al.,* 2010) and was considered optimal for efficient carbohydrate utilization over dark/light cycle (Smith and Stitt, 2007).

Plants were seen to dynamically assess the starch content and the expected time to dawn, divide the two quantities and accordingly compute the appropriate starch degradation rate. From these observations, Scialdone *et al.,* (2013) believed that plants had the ability for arithmetic computation. In this way, they could ensure complete utilization of available starch reserves by dawn despite variations in both the starch content at the onset of darkness and the subsequent duration of darkness. In case of an early night, they were able to adjust the degradation rate, which was further verified by exposing plants to darkness for imposing unexpected night periods. They found that the decline in starch content was linear with time in all cases but the slope of degradation varied; moreover, in all cases the starch reserves were almost exhausted by dawn.

Plant's ability to compute starch degradation rate based on starch availability and time to dawn clearly indicated the problem solving capacity of plants. In order to assess the role of starch degradation apparatus on controlling starch degradation rate, six mutant plants each lacking an enzymes involved in starch degradation were used and an unexpectedly early night was imposed on them., The mutants chosen for the study included *lsf1*, *sex4*, *bam3*, *bam4*, *isa3*, *pwd* with specific roles in starch degradation such that *LSF1* is required for proper starch degradation wherein it acts as a β-Amylase-Binding Scaffold; *STARCH EXCESS4* (*SEX4*) phosphorylates the C6 position of the glucosyl residues; proteins *BAM3* and *BAM4* possess a typical glucosyl hydrolase domain and is thought to act as a regulator of starch degradation; *ISOAMYLASE3* removes external chains of amylopectin that have been shortened by β-amylolysis; and *PHOSPHOGLUCAN WATER DIKINASE* (*PWD*) is involved in the phosphorylation of the glucosyl residues of starch at the C3 position.

The six mutants used for the study included most of the currently known components of the chloroplastic starch degradation apparatus. They changed the night period and studied the starch content at the end of the dark period. All the mutants had higher starch content than wild type at the end of the dark period. They also noted that the rate of starch degradation was lower during an unexpected early night (starting 8 h after dawn) than a normal night (starting 12 h after dawn). To quantify the adjustment of starch degradation rates in the mutants, the ratio R (ratio between the expected lengths of early and normal night periods) between degradation rates (normalized by the respective end-of-light period starch content) during normal and early night was calculated. Most of the mutants had an R value similar to the wild type but for the *pwd*, mutant, the R value was found to be different from that of the wild type. This mutant alone could not adjust the rate of starch degradation when an unexpected early night was imposed. From this it was evident that *PWD* was the key enzyme that regulated the rate of starch degradation based on the duration of night period. *PWD* controls phosphorylation/dephosphorylation cycles of starch degradation by acting on glucosyl residues within starch polymers. The phosphorylation/dephosphorylation process modifies the granular surface of starch grains and also stores information about the starch content. This was also corroborated by changes in the level of phosphate per unit of starch over the light/dark period.

Similarities between plant neurobiology with animal/human neurobiology

Scientists have found several similarities between plant cells and neurons, especially in complex information processing, where plants adopt not only action potentials but also synaptic modes of cell–cell communication. Vesicular trafficking and transport of auxin in plants resembles neuro-transmitters in animals (Baluska, 2010).

Resemblance between root morphogenesis to neuron extending axon

Neuron morphogenesis indicates that both microtubules and actin filaments have fundamental roles in the development of axons. The two cytoskeletal systems have mostly distinct roles in axonal biology, but cooperate in order to generate a fully functional axon. During the development of neuron, the neural growth cone extends to reach its target during which there is accumulation and rearrangement of actin filaments. This can be compared with root hair formation in plants where the root hair outgrowth is associated with microtubule (MT) and actin filament reorientations (Emons and Derksen, 1986; Baluška *et al.*, 2000). During the initial phase of root hair formation, an outgrowing bulge is formed by the accumulation of actin filaments. Further rearrangement and accumulation of actin filaments allows the elongation of root hairs. Moreover, they also act as a structural scaffold for diverse signaling complexes in eukaryotes (Juliano, 2002). Recent data from plants support the concept that dynamic actin cytoskeleton is closely linked to signaling cascades initiated at the plasma membrane (Meagher *et al.*, 1999).

Vesicular trafficking

Vesicular trafficking involves all the logistics mediated by vesicles as well as the organelles that send and receive vesicles. In plants, vesicular trafficking has two pathways-(i) secretary pathway which starts in the endoplasmic reticulum where proteins are packed in vesicles and transported in to Golgi apparatus where it travels through trans-Golgi network and reaches the plasma membrane and releases proteins by exocytosis; and (ii) importing pathway which starts at the plasma membrane where the vesicles are formed by invaginations of plasma membrane. These vesicles are transported to endosomes which are converted to lysosomes, which degrade the proteins (Uemura *et al.*, 2014).

In animal neurons, vesicular trafficking pathway is referred to as axon transport. The two pathways are orthograde or anterograde transport, where proteins, neurotransmitters, RNA *etc.* produced by cell body are transported from cell body to axon terminal. In retrograde transport, material transport is from axon terminal to cell body.

Resemblance between vesicular transport of auxin and neurotransmitters

Recent studies suggest that root apices along with their primary role in the uptake of nutrients, also support neuronal-like activities based on plant synapses. These synapses transport auxin suggesting that it is a plant-specific neurotransmitter-like moiety which resembles neurotransmitter in animals. Baluska *et al.*, (2004) reported that auxin release and neurotransmitter release are similar. Auxin and neurotransmitters are secreted *via* exocytosis wherein, auxin is loaded into vesicles by PIN1 transporter and neurotransmitters such as glutamate are loaded into vesicles by glutamate transporters. The loading force in both cases is provided by the action of vesicle membrane bound H^+ ATPases (Baluska *et al.*, 2008).

Neurotransmitters from plants

The following are a few neurotransmitters identified in plants which regulate physiological processes (Roschina, 2010).

Table 1. Plant neurotransmitters and their possible roles

Substance	Process (function)
Acetylcholine	Germination, Flowering, Leaf movement, Photomorphogenesis, Protoplast swelling, Ion permeability
Dopamine	Protective role, Reproductive organogenesis, Ion permeability
Noradrenaline	Adaptation to environmental changes, Flowering, Morphogenesis, Ion permeability
Adrenaline	Protective role, Somatic embryogenesis, Reproductive organogenesis, Flowering, Ion permeability
Serotonin	Protective role, Adaptation to environmental changes, Flowering, Morphogenesis, Ion permeability
Histamine	Regulation of growth and development at stress

Cell- cell adhesion domains or synapses

Plant synapses are defined as 'actin-based asymmetric adhesion domains assembled at cellular end-poles (cross-walls) between adjacent plant cells for rapid cell-to-cell communication which is accomplished by vesicle trafficking' (Baluska *et al.,* 2004).

Vascular strands as plant neurons

Vascular system of plants show resemblance to nervous system of animals wherein the strands look like "nerves" and the other components form the endoskeleton. Up to 50% of the root diameter is constituted by vascular tissue, and its strands are supported by numerous 'nursery' cells forming the vascular cylinder (Sachs, 2000). Stelar tissues in roots are completely enclosed by meristematic pericycle and protective endodermis and they are very active in transcellular auxin transport while pericycle cells initiate lateral root formation. Leaves contain thin single strands which are joined together to form vascular bundles which extends from the stem to the root.

The phloem which connects the shoot and root apices can be considered as a supra cellular axon-like 'channel'. It is reported that phloem cells can transmit action-potential-driven electric signals (Mancuso, 1999) while xylem tubes are specialized for transmission of hydraulic signals. These signals are transmitted as waves which are induced and driven by changes in hydrostatic pressure (Mancuso, 1999).

Concept of plant brain

The concept of 'Plant Brain' was first proposed by Charles Darwin in 1880. According to him "plant brains are localized within root apices at the anterior pole of the plant body". In shoots, it is seen that cell division and cell elongation occur concomitantly while in the root, elongation is more rapid than shoot elongation, as a result there will be a 'no cell division zone' in the region of rapid elongation. This is a transition zone between cell division and elongation and is interpolated between different regions have a unique cyto-architecture, with centralized nuclei surrounded by perinuclear microtubules radiating towards the cell periphery (Baluska *et al.,* 2001). This specific configuration of cells is suited for perception and transmission of signals to and from the nuclei. It is seen that these cells are not involved in either cell division or elongation and their only function is perception of environmental signals and processing of developmental cues. Moreover, root apices also form the site for the perception of low-temperature and drought signals and transmit this information to aerial plant parts and shoot apices (Blake and Ferrell, 1977). When herbivores attack above-ground aerial shoots, it is seen that root apices perceive these signals and activate the production of VOCs and take part in plant-to-plant communication (Chamberlain *et al.,* 2001).

Each root apex is proposed to possess brain-like units of the nervous system of plants. The number of root apices in the plant body is high, and all 'brain units' are interconnected via vascular strands (plant neurons) with their polarly-transported auxin, forming a serial /parallel neuronal system of plants (Baluska *et al.,* 2006).

Sensory zones in root apex

There are two clearly defined sensory zones in the root apex: root cap covering the meristem and transition zone interpolated between meristem and elongation region. Both these sensory zones receive diverse signals and the output is a differential switch-like onset of rapid cell elongation, resulting in either straight growth (when all postmitotic cells start their rapid cell elongation simultaneously) or rapid turnings of root apex. The transition zone is flooded with sucrose, which allows energy demanding 'brain-like' information processing in cells of transition zone (Baluska *et al.,* 2006).

Applications

Biorobotics-The plantoid

Plantoid is a robot which acts, looks and grows like plants. They mimic the soil penetration, exploration and monitoring abilities of roots. Plantoids have applications in detection of pollutants and heavy metals, nutrient sensing, water searching and soil monitoring *etc.* (Cordis, 2014).

VOCs in crop improvement

VOCs such as methyl jasmonate, methyl salicylate, green leaf volatiles etc. have been used to make the plant withstand abiotic and biotic stresses. VOCs such as citral, carvacrol, trans-2-hexenal have been reported to inhibit growth and development of pathogens (Brilli *et al.,* 2019).

Terra farming

Terra farming is making the environment of other planets such as Venus similar to Earth. As a preliminary step, bacteria and plants are grown by providing environmental conditions similar to that of Venus on Earth, based on their neurobiological response capacities.

Conclusion

Plant neurobiology deals with signaling and behavior of plants. They can obtain information from environment and are able to respond to these stimuli. They can perceive signals from environment, process the information and even communicate with their neighbours, by the production of wide array of volatile organic compounds (VOCs). Contrary to our earlier perception, they also seem to have a rich social life.

They seem to have their own likes and dislikes and show cooperative or antagonistic behavior towards other components of the ecosystem. Finally, as suggested by Darwin, plants have root transition zones which resemble a diffuse nervous system and may be categorized as the "plant brain". Hence our outlook on plants and their interaction needs to be redefined.

References

Baluska, F. 2010. Recent surprising similarities between plant cells and neurons. *Plant Signal. Behav*. 5(2):87-89.

Baluska, F., Mancuso, S., Volkmann, D., and Barlow, P. 2004. Root apices as plant command centres: the unique 'brain-like' status of the root apex transition zone. *Biologia* 13: 7- 19.

Baluska, F., Salaj, J., Mathur, J., Braun, M., Jasper, F., Barlow, P. W., and Volkman, D. 2000. Root hair formation: F-Actin-dependent tip growth is initiated by local assembly of profilin-supported F-Actin meshworks accumulated within expansin-enriched bulges. *Dev. Biol*. 227:618-632.

Baluska, F., Schlicht, M., Volkmann, D., and Mancuso, S. 2008. Vesicular secretion of auxin. *Plant Signal. Behav*. 3(4):254-256.

Baluska, F., Volkman, D., and Barlow, P.W. 2001. A polarity crossroad in the transition growth zone of maize root apices: cytoskeletal and developmental implications. *J. Plant Growth Regul*. 20: 170-181.

Baluska, F., Volkmann, D., Hlavacka, A., Mancuso, S., and Barlow, P.W. 2006. *Communication in Plants* [e-book]. Available: https://doi.org/10.1007/978-3-540-28516-8_2 [30 April 2019].

Blake, J. and Ferrell, W. K. 1977. The association between soil and xylem water potential, leaf resistance, and abscisic acid content in droughted seedlings of Douglas-fir (*Pseudotsuga menziesii*). *Physiol. Plant*. 39: 106–109.

Bose, J. C. 1907. *Comparative Electro-physiology*. Longman, Green and Co., London.

Brilli, F., Loreto, F., and Baccelli, I. 2019. Exploiting plant volatile organic compounds (VOCs) in agriculture to improve sustainable defense strategies and productivity of crop. *Front. Plant Sci*. 10: 264-272.

Chamberlain, K., Guerrieri, E., Pennacchio, J., Pettersson, J., Pickett, P. A., Poppy, G. M., Powell, W., Wadhams, L. J., and Woodcock, C. M. 2001. *Biochem. Syst. Ecol*. 29: 1063-1074.

Cordis. 2014. Plantoid: building a robot to mimic plants [on-line]. Available: https://phys.org/news/2014-05-plantoid-robot-mimic.html [25 Nov. 2019].

Darwin, C. H. 1880. *The Power of Movements in Plants*. John Murray, London, 750p.

Emons, A. M. C. and Derksen, J. 1986. Microfibrils, microtubules and microfilaments of the trichoblast of *Equisetum hyemale*. *Acta Botanica Neerlandica*, 35: 311-320.

Farmer, E. E. and Ryan, C. A. 1990. Interplant communication: airborne methyl jasmonate induces synthesis of proteinase inhibitors in plant leaves. *Proc. Natl. Acad. Sci*. U.S.A. 87(19):7713-716.

Gibon, Y., Blasing, O. E., Palacios, N., Pankovic, D., Hendriks, J. H., and Fisahn, J. 2004. Adjustment of diurnal starch turnover to short days: depletion of sugar during the night leads to a temporary inhibition of carbohydrate utilization, accumulation of sugars and post-translational activation of ADP-glucose pyrophosphorylase in the following light period. *Plant J*. 39: 847-62.

Graf, A., Schlereth, A., Stitt, M. and Smith, A. M. 2010. Circadian control of carbohydrate availability for growth in Arabidopsis plants at night. *Proc. Natl. Acad. Sci.* U.S.A. 107: 58-63.

Haberlandt, G. 1884. *Physiological Plant Anatomy*. Macmillan, London, 763p.

Hare, J. D. Ecological role of volatiles produced by plants in response to damage by herbivorous insects. 2011. *Annu. Rev. Entomol.* 56: 161-80.

Hirokazu, U., Kikuta, Y., and Matsuda, K. 2012. Plant communication mediated by individual or blended VOCs? *Plant Signal. Behav.* 7(2): 222-226.

Houwink, A.L. 1935. The conduction of excitation in *Mimosa pudica. Rec. Tray. Bot. Neerl.* 32: 51-91.

Juliano, R. L. 2002. Signal transduction by cell adhesion receptors and the cytoskeleton: functions of integrins, cadherins, selectins, and immunoglobulin-superfamily members. *Annu. Rev. Pharmacol. Toxicol.* 42: 283–323.

Koptur, S. 1992. *Insect- Plant Interactions*: *Vol. 4. Extra-floral nectary-mediated interactions between insects and plants*. CRC Press, Boca Raton, 50p.

Lake, J. A., Quick, W. P., Beerling, D. J., and Woodward, F. I. 2001. Signals from mature to new leaves. *Nat.* 411: 154-155.

Lucht, J. M. 2002. Pathogen stress increases somatic recombination frequency in Arabidopsis. *Nat. Genet.* 30: 311-314.

Mancuso, S. 1999. Hydraulic and electrical transmission of wound-induced signals in *Vitis vinifera. Aust. J. Plant Physiol.* 26: 55-61.

Meagher, R. B., McKinney, E. C., and Kandasamy M. K. 1999. Isovariant dynamics expand and buffer the responses of complex systems: the diverse plant actin gene family. *Plant Cell* 11: 995–1006.

Molinier, J., Ries, G., Zipfel, C., and Hohn, B. 2006. Transgeneration memory of stress in plants. *Nat.* 442: 1046-1049.

Opritov, V. A. and Pyatygin, S. S. 1989. Evidence for coupling of the action potential generation with the electrogenic cornponent of the resting potential in *Cucurbita pepo* L. Stern excitable cells. *Biochem. Physiol. Pflanzen* 184: 447-451.

Press, M. C. and Phoenix, G. K. 2005. Impacts of parasitic plants on natural communities. *New Phytol.* 166(3): 737-751.

Ries, G., Heller, W., Puchta, H., Sandermann, H., Seidlitz, H. K., and Hohn, B. 2000. Elevated UV-B radiation reduces genome stability in plants. *Nat.* 406: 98-101.

Roshchina, V. V. 2010. Evolutionary Considerations of Neurotransmitters in Microbial, Plant, and Animal Cells. In: Lyte M., Freestone P. (eds) *Microbial Endocrinology*. Springer, New York, NY.

Runyon, J. B., Mescher, M. C., Consuelo, M., and Moraes, D. 2006. Volatile chemical cues guide host location and host selection by parasitic plants. *Sci.*313: 1964-1967.

Sachs, T. 2004. Self-organization of tree form: a model for complex social systems. *J. Theor. Biol.* 230:197-202.

Scialdone, A., Mugford, S. T., Feike, D., Skeffington, A., Borrill, P., and Howard, M. 2013. Arabidopsis plants perform arithmetic division to prevent starvation at night. *eLife* [e-journal] 2(1). Available:https://elifesciences.org/articles/00669 [10April2019].

Sibaoka, T. 1969. Physiology of rapid movements in higher plants. *Ann. Rev.* Plant *Physiol.* 20:165-84.

Smith, A. M. and Stitt, M. 2007. Coordination of carbon supply and plant growth. *Plant Cell Environ.* 30: 26-49.

Stahlberg, R. 2006. Historical overview on plant neurobiology. *Plant Signal. Behav.* 1(1): 6- 8.

Uemura, T., Suda, Y., Ueda, T., and Nakano, A. 2014. Dynamic Behavior of the trans-Golgi Network in Root Tissues of Arabidopsis Revealed by Super-Resolution Live Imaging. *Plant Cell Physiol.* 55: 694-703.

Vickers, C. E., Gershenzon, J., Lerdau, M. T., and Loreto, F. 2009. A unified mechanism of action for volatile isoprenoids in plant abiotic stress. *Nat. Chem. Biol.* 5: 283-91.

Yazdanbakhsh, N., Sulpice, R., Graf, A., Stitt, M., and Fisahn, J. 2011. Circadian control of root elongation and C partitioning in *Arabidopsis thaliana. Plant Cell Environ.* 34: 877-94.

Yoder, J.I. 1997. A species-specific recognition system directs haustorium development in the parasitic plant *Triphysaria* (*Scrophulariaceae*). *Planta* 202(4):407-413.